AUTOCRINE AND PARACRINE MECHANISMS IN REPRODUCTIVE ENDOCRINOLOGY

REPRODUCTIVE BIOLOGY

Series Editor: Sheldon J. Segal
The Rockefeller Foundation
New York, New York

THE ANTIPROGESTIN STEROID RU 486 AND HUMAN FERTILITY CONTROL
Edited by Etienne-Emile Baulieu and Sheldon J. Segal

AUTOCRINE AND PARACRINE MECHANISMS IN REPRODUCTIVE ENDOCRINOLOGY
Edited by Lewis C. Krey, Bela J. Gulyas, and John A. McCracken

CONTRACEPTIVE STEROIDS: Pharmacology and Safety
Edited by A. T. Gregoire and Richard P. Blye

DEMOGRAPHIC AND PROGRAMMATIC CONSEQUENCES OF CONTRACEPTIVE INNOVATIONS
Edited by Sheldon J. Segal, Amy O. Tsui, and Susan M. Rogers

GENETIC MARKERS OF SEX DIFFERENTIATION
Edited by Florence P. Haseltine, Michael E. McClure, and Ellen H. Goldberg

GOSSYPOL: A Potential Contraceptive for Men
Edited by Sheldon J. Segal

IMMUNOLOGICAL APPROACHES TO CONTRACEPTION AND PROMOTION OF FERTILITY
Edited by G. P. Talwar

A Continuation Order Plan is available for this series. A continuation order will bring delivery of each new volume immediately upon publication. Volumes are billed only upon actual shipment. For further information please contact the publisher.

AUTOCRINE AND PARACRINE MECHANISMS IN REPRODUCTIVE ENDOCRINOLOGY

Edited by

Lewis C. Krey

Rockefeller University
New York, New York

Bela J. Gulyas

National Institutes of Health
Bethesda, Maryland

and

John A. McCracken

Worcester Foundation for Experimental Biology
Shrewsbury, Massachusetts

PLENUM PRESS • NEW YORK AND LONDON

Library of Congress Cataloging-in-Publication Data

Reproductive Endocrinology Study Section Workshop on Autocrine and Paracrine Mechanisms in Reproductive Endocrinology (1988 : Shrewsbury, Mass.)
Autocrine and paracrine mechanisms in reproductive endocrinology / edited by Lewis C. Krey, Bela J. Gulyas, and John A. McCracken.
p. cm. -- (Reproductive biology.)
"Proceedings of a Reproductive Endocrinology Study Section Workshop on Autocrine and Paracrine Mechanisms in Reproductive Endocrinology, held October 14-16, 1988, in Shrewsbury Massachusetts"--T.p. verso.
Organized by the members and Executive Secretary of the Reproductive Endocrinology Study Section, Division of Research Grants, National Institutes of Health"--Pref.
Includes bibliographical references.

DOI 10.1007/978-1-4684-5751-3
1. Human reproduction--Endocrine aspects--Congresses. 2. Paracrine mechanisms--Congresses. 3. Autocrine mechanisms--Congresses. 4. Growth factors--Physiological effect--Congresses. I. Krey, Lewis C. II. Gulyas, Bela J. III. McCracken, John A. IV. National Institutes of Health (U.S.). Reproductive Endocrinology Study Section. V. Title. VI. Series.
[DNLM: 1. Endocrine Glands--physiology--congresses. 2. Growth Substances--physiology--congresses. 3. Reproduction--congresses. WQ 205 R42871 1988]
QP252.R47 1988
612.6--dc20
DNLM/DLC
for Library of Congress 90-6703
CIP

Proceedings of a Reproductive Endocrinology Study Section Workshop on Autocrine and Paracrine Mechanisms in Reproductive Endocrinology, held October 14–16, 1988, in Shrewsbury Massachusetts

MyCopy version of the original edition 1989

A Division of Plenum Publishing Corporation
233 Spring Street, New York, N.Y. 10013

PREFACE

There is a provision in the charter of each Study Section of the Division of Research Grants at the National Institutes of Health that stipulates that "workshops" are to be held periodically to aid Study Section members in their appraisals of recent developments in their fields and to identify future challenges worthy of investigation. The Reproductive Endocrinology Study Section was established on December 13, 1985 to review research grants and research training activities relating to reproductive endocrinology, including aspects of management of reproductive endocrine disorders and hormonal imbalances as related to infertility and during pregnancy and puberty, breast cancer and prostate cancer. It held its first workshop, entitled, "Autocrine and Paracrine Mechanisms in Reproductive Endocrinology," in October, 1988 in Shrewsbury, MA at The Worcester Foundation for Experimental Biology. The proceedings of this workshop, which are detailed herein, reflect the fact that autocrine and paracrine interactions are rapidly being accepted as an exciting area of research by scientists investigating the physiological and biochemical mechanisms of hormone action in the male and female reproductive systems. The material covered is novel and wide-ranging, extending from theoretical considerations of mechanisms of growth factor action and the role of cell cycle stage in determining hormone action to investigations of autocrine and paracrine interactions during development to discussions of the potential clinical ramifications of the basic research findings. Such an extensive inventory is necessary for two reasons. First, paracrine and autocrine interactions appear to be involved in a broad spectrum of physiological functions in many reproductive tissues. Secondly, these presentations constitute one of the first focussed examinations of research on paracrine and autocrine phenomena in endocrinology.

The workshop on "Autocrine and Paracrine Mechanisms in Reproductive Endocrinology" was organized by the members and Executive Secretary of the Reproductive Endocrinology Study Section, Division of Research Grants, National Institutes of Health. The members of the Organizing Committee included Drs. Bela J. Gulyas (Executive Secretary), Lewis C. Krey, member of the Reproductive Endocrinology Study Section, and John A. McCracken, former member of the Study Section. Members of the Program Committee included Drs. Gulyas, Abubaker A. Shaikh (Incoming Executive Secretary), Robert M. Brenner, Neil J. MacLusky, Kamran S. Moghissi, Richard C. Parker, Mary Ruh and Theresa M. Siler-Khodr. The workshop was supported by the Division of Research Grants, with the assistance of the National Cancer Institute, the National Institute of Diabetes and Digestive and Kidney Diseases and The Worcester Foundation for Experimental Biology. The Study Section is deeply indebted to the following individuals who made the workshop possible: Drs. Raymond E. Bahor and Faye J. Calhoun of the

Division of Research Grants; Drs. Andrew Chiarodo and Colette S. Freeman of the National Cancer Institute; and Drs. Francisco O. Calvo and Robert A. Tolman of the National Institute of Diabetes and Digestive and Kidney Diseases.

In addition, the members of the Study Section would like to thank the following individuals for their help in making arrangements for the workshop and in the preparation of its proceedings for publication: Mary Berlin (Grants Technical Assistant), who played a major role in coordinating the organization of the Workshop; Wendy Roine, who expertly transposed the manuscripts into a camera-ready format for publication; and Greg Safford and Melanie Yelity at Plenum Press who provided counsel during the preparation of the book.

Considering the focus of the research presented and discussed, it was especially fitting that the workshop was held at The Worcester Foundation for Experimental Biology in conjunction with its awarding of the Gregory Pincus Award and Memorial Lectureship to Dr. Stanley Cohen for his pioneering work on epidermal growth factor. The members of the Study Section congratulate Stan for this prestigious award and would like to thank Dr. Tibor Pedersen, President and Scientific Director, for inviting them to share in this ceremony and for hosting the workshop and Dr. John A. McCracken, Senior Scientist, for making the local arrangements.

In view of the relevance of Dr. Cohen's research accomplishments to our current understanding of autocrine and paracrine phenomena, we have asked him to contribute some brief reflections on the growth of research on "his factor" and to provide some challenges for future work. He has generously consented with the following remarks:

"Epidermal growth factor (EGF) is now almost 30 years old. While the importance of 'classical' hormones in the control of growth and development had been long recognized, interest in 'factors,' whose biological activities were usually assayed with crude extracts, was slow to develop. The isolation and determination of the primary amino acid sequence of these 'factors' made them real, not only to the biological community but to the chemically-oriented community as well.

It could hardly have been anticipated that exploration of the initial discovery that extracts of male mouse submaxillary glands induced precocious eyelid opening when injected into new-born animals would yield such a rich harvest: the isolation and primary structure of EGF; the recognition that eyelid-opening was due to the direct effect of EGF on epidermal growth; the demonstration that peptide hormones could be internalized together wth their receptors by absorptive pinocytosis; a new general mechanism for hormone-receptor interactions - that receptors function as ligand-activated tyrosine kinases; the detection of families of EGF-related proteins some of which appear to interact with the EGF receptor; the elucidation of the amino acid sequence of the EGF-receptor and the discovery that the erb-β transforming gene of avian erythroblastosis probably is derived from the avian EGF receptor; and finally, the elucidation of nucleotide sequence of the cDNA for prepro-EGF, that predicts a very large, membrane spanning protein precursor that may itself be a receptor for an unknown ligand.

Despite these clear advances, which have been made in many laboratories around the world, answers to many vital questions remain elusive. How does the binding of EGF to the extracellular portion of the EGF receptor activate its intracellular kinase activity ? Are there specific cellular proteins whose functions are altered by tyrosine phosphorylation ? Is the intracellular translocation of tyrosine kinases of physiological significance ? Is it possible that autophosphorylated receptor/kinases or related oncogene proteins serve some still unidentified regulatory role ? What are the mechanisms for sending stimulatory or inhibitory signals to the cell nucleus ? What is the normal physiological role of EGF during development and homeostasis ? Does the large membrane-spanning prepro-EGF molecule have a biological function other than as a precursor for mature EGF ? The answers to these and a host of other questions must be found before we can fully comprehend this important regulatory system.

This workshop is testimony that many more important insights about EGF and other 'factors' are to come in the future. Hopefully, with this understanding will come medical applications to benefit all of us."

The members of the Reproductive Endocrinology Study Section share Dr. Cohen's enthusiasm for research investigating the role of growth factors during development and homeostasis. We hope that the material detailed in this volume concerning the roles these factors may play within the breast and reproductive tract, as well as the studies reporting on the other exciting aspects of autocrine and paracrine phenomena, will serve as guidelines for future research in this area and contribute to our understanding of the physiology and pathologies of the male and female reproductive systems.

Lewis C. Krey
Bela J. Gulyas
John A. McCracken

CONTENTS

LIPOCORTINS AND RELATED PROTEINS MAY BE INVOLVED IN INTRACELLULAR SIGNAL TRANSDUCTION

Harry T. Haigler

Department of Physiology and Biophysics
University of California
Irvine, CA 92717

Recent studies have defined a family of structurally related proteins which bind to certain phospholipids in Ca^{2+}-dependent manner (Crompton *et al.*, 1988). Several names have been proposed for this family including annexins, lipocortins, and calpactins. Although the biological function is not known for any of these proteins, they are attracting intensive investigation because of their potential involvement in Ca^{2+}-mediated stimulus-response coupling.

Annexins are structurally related proteins that bind to certain phospholipids in a Ca^{2+}-dependent manner. They were identified by investigators with apparently unrelated goals of identifying mediators of exocytosis, substrates for protein-tyrosine kinases, inhibitors of phospholipase A_2, inhibitors of coagulation, and components of the cytoskeleton. Complete or partial sequence analysis has established that there are at least nine distinct annexins. One has an apparent M_r of 68,000 while the others have an apparent M_r of approximately 35,000. Certain of the proteins are known by more than one name because their identities were not recognized until after their amino acid sequences were determined. The seven annexins of known complete sequences are listed in Table 1.

Each annexin has two domains: a small amino-terminal domain and a core domain that is formed by either a 4-fold or 8-fold repeat of a conserved segment. The conserved segment contains approximately 70 amino acids and there is 25-35% sequence identity between repeats within an individual protein. The structures of four of the annexins with a four-fold repeat are compared in Figure 1. Different annexins share little sequence similarity in the amino-terminal domain but have 40-60% sequence identity in the core domain. The Ca^{2+} and phospholipid binding sites are located in the conserved core domain (Glenney, 1986; Schlaepfer and Haigler, 1987). Each conserved 70-amino acid segment probably forms a Ca^{2+} binding site (Schlaepfer *et al.*, 1987). The core domains of the 35-kDa annexins probably arose as a series of two sequential gene duplications of an ancestral gene encoding a Ca^{2+} binding protein. An additional duplication may have formed the 68-kDa annexin. The individual annexins probably arose after these gene duplications occurred, because there is more structural similarity between different annexins within the core domains than there is between the repeats within a particular protein.

Table I. Nomenclature of Annexins

lipocortin	1 (1)	2 (2)	3 (3)	4 (3)	5 (3)	6 (3)	
kinase substrate	p35 (4,5)	p36 (6)					
calpactin	II (7)	I (7)					
"protein"		I (8)		II (9)		III (8)	
chromobindin	9 or 6 (10)	8 (10)		4 (10)	5 or 7 (10)	20 (10)	
calelectrin				32-kDa (11)		67-kDa (11)	
calcimedin			35 (12)			68 (12)	
endonexin				I (13)	II (14)		
other	GIF (15)				PAP (16,17,18)	p68 (19)	synexin (20)
apparent Mr by PAGE	35,000	36,000	35,000	33,000	33,000	68,000	47,000
apparent pI	6.5-6.8	7.4-7.8		5.4-5.8	4.9-5.2	5.7-6.4	7.0

Each of the six columns represents a single protein: e.g. lipocortin 1, p35, and calpactin II are identical proteins. PAP-placental anticoagulant protein. GIF - glycosylation-inhibiting factor. References are given in parenthesis and focus on the most recent studies and not on initial identification of the proteins. The references cited include: 1 (Wallner et al., 1986); 2 (Huang et al., 1986; 3 (Pepinsky et al., 1988); 4 (De et al., 1986); 5 (Haigler et al., 1987); 6 (Saris et al., 1986); 7 (Glenney, 1986); 8 (Shadle et al., 1985); 9 (Weber et al., 1987); 10 (Creutz et al., 1987); 11 (Sudhof et al., 1988); 12 (Smith and Dedman, 1986); 13 (Geisow et al., 1986); 14 (Kaplan et al., 1988); 15 (Iwata and Ishizaka, 1987); 16 (Funakoshi et al., 1987a); 17 (Iwasaki et al., 1987); 18 (Funakoshi et al., 1987b); 19 (Crompton et al., 1988); 20 (Burns et al., 1989).

```
Amino     ┌  I. Endonexin II                        MAQVLRGTVTDFPGFDERA  19
Terminal  │ II. Protein II                              AAKG---KAAS--NAAE  17
Domain    │III. Calpactin I   MSTVHEILCKLSLEGDHSTPPSAYG.S-KAYTN--AER  37
          └ IV. Lipocortin I  MAMVSEFL.K..QAWFIENEE-EYVQ--KSSK-GPGS-VSPYPTFNPSS  46

          ┌  I.  20  DAETLRKA.MKGLGTDEESILTLLTSRSNAQRQEISAAFKTLFGRDLLDDLKSELTGKFEKLIVALMKPSRLY   91
First     │ II.  18  --Q-----.--------DA-ISV-AY--T------RT-Y-STI------------S-N--QV-LGM-T-TV--   89
Repeat    │III.  38  --LNIET-.I-TK-V--VT-VNI--N-------D-AF-YQRRTKKE-ASA---A-S-HL-TV-LG-L-TPAQ-  109
          └ IV.  47  -VAA-H--.IMVK-V--AT-IDI--K-N-----Q-K--YLQET-KP-DET--KA---HL-EVVL--L-TPAQF  118

          ┌  I.  92  DAYELKHA.LKGAGTNEKVLTEIIASRTPEELRAIKQVYEEEYGSSLEDDVVGDTSGYYQRMLVVLLQANRDPDAG.IDEAQVEQ  174
Second    │ II.  90  -VQ--RR-.M-----D-GC-I--L-------I-R-N-T-QLQ--R-----IRS---FMF--V--S-SAGG--E.GNYL-D-L-R-  172
Repeat    │III. 110  --S---AS.M--L--D-DS-I---C---NQ--QE-NR--K-M-KTD--K-IIS----DFRKLM-A-AKGR-AE-GSV--YELID-  193
          └ IV. 119  --D--RA-.M--L--D-DT-I--L----NK-I-D-NR--R--LKRD-AK-ITS----DFRNA-LS-AKGD-SE-F-VNED.LADS  201

          ┌  I. 175  DAQALFQAGELKWGTDEEKFITIFGTRSVSHLRKVFDKYMTISGFQIEETIDRETSGNLEQLLLAVVKSIRSIPAYL  251
Third     │ II. 173  ---D-YE---K------V--L-VLCS-NRN--LH---E-KR--QKD--QS-KS----SF-DA---I--CM-NKS--F  249
Repeat    │III. 194  --RD-YD--VKRK---VP-W-S-MTE---P--Q----R-KSY-PYDML-S-RK-VK-D--NAF-NL-QC-QNK-L-F  270
          └ IV. 202  --R--YE---RRK---VNV-N--LT---YPQ--R--Q--TKY-KHDMNKVL-L-LK-DI-KC-T-I--CAT-K--FF  278

          ┌  I. 253  .AETLYYA.MKGAGTDDHTLIRVMVSRSEIDLFNIRKEFRKNFATSLYSMIKGDTSGDYKKALLLLCGEDD  320.
Fourth    │ II. 250  .--R--KS.---L----N---------A---MMD--AN-KRLYGK----F---------R-V--I---G--  318.
Repeat    │III. 271  .-DR--DS.---K--R-KV---I------V-MLK--S--KRKYGK---YY-QQ--K---Q----Y---G--  339.
          └ IV. 279  .--K-HQ-.---V--RHKA---I--------MND-KAFYQ-MYGI--CQA-LDE-K---E-I-VA---GN  346.

conserved residues   DA  L  A    GT           R       I   Y      L       G       L
```

Figure 1. Comparison of the amino acid sequence of four annexins. The primary amino acid sequence of human endonexin II (Geisow *et al.*, 1986), bovine protein II (DiRosa *et al.*, 1984), human calpactin I or p36 or lipocortin 2 (Cheng and Chen, 1981), and human lipocortin 1 (Burns *et al.*, 1989) as represented by the single letter amino acid code are aligned with the inclusion of gaps (".") to emphasize the similarity between the proteins. Residues that are identical to the corresponding residues in endonexin II are indicated ("-").
All four proteins contain an amino terminal domain with limited sequence similarity and a conserved 4-fold internal repeat. Highly conserved residues in therepeats are shown at their corresponding positions on the last line: invariant residues are underlined. The following residues are phosphorylated by protein kinase C: ser-6 of protein II (DiRosa *et al.*, 1984), ser-26 of calpactin I (Pfaffle *et al.*, 1988), and thr-24, ser-27, and ser-28 of lipocortin I (Pepinsky *et al.*, 1988). Tyr-21 of lipocortin I (Crompton *et al.*, 1988 (Haigler *et al.*, 1988) and tyr-24 of calpactin I (Weber *et al.*, 1987) are phosphorylated by the EGF receptor/kinase and by pp60^{src}, respectively.

```
Amino     [rat lipocortin V                ..--....... S  11
Terminal  |human endonexin II              MAQVLRGTVTDFP  13
Domain    [chicken anchorin CII            ..KYT.....A.S  19

First     [rat     ...G....V....-........D...N...A.......Q.AEE.........VN.M................  83
Repeat    |human   GFDERADAETLRKA-MKGLGTDEESILTLLTSRSNAQRQEISAAFKTLFGRDLLDDLKSELTGKFEKLIVALM  85
          [chick   P..A.....A....-...M.....T..KI....N.......AS.........V...........T.M.S..  91

Second    [rat     ..............-......D....................A.......N.........................TA..-- 166
Repeat    |human   KPSRLYDAYELKHA-LKGAGTNEKVLTEIIASRTPEELRAIKQVYEEEYGSSLEDDVVGDTSGYYQRMLVVLLQANRDPDAGID-- 168
          [chick   R.A.IF..HA....-I............L....LLKC.ILNRFNMQ..EAN.GRNKITGRRHQAIFRDCWWSCCRQIEIPDGRVE 176

Third     [rat     ...........................L..................................N.........  245
Repeat    |human   EAQVEQDAQALFQAGELKWGTDEEKFITIFGTRSVSHLRKVFDKYMTISGFQIEETIDRETSGNLEQLLLAVVKSI  247
          [chick   ..L..K...V..R..........T....L........R......................D..K.......C.  255

Fourth    [rat     ..............-.............I.........................................GE.D*  319
Repeat    |human   RSIPAYLAETLYYA-MKGAGTDDHTLIRVMVSRSEIDLFNIRKEFRKNFATSLYSMIKGDTSGDYKKALLLLCGEDD*  320
          [chick   ..V...F......S-.......D.............LD..H.......K...Q..Q.......R........G..E*  329
```

Figure 2. Comparison of anchorin CII sequence with annexins. The primary amino acid sequences of rat lipocortin V (Pepinsky *et al.*, 1988), human endonexin II (Cheng and Chen, 1981; funakoshi *et al.*, 1987a,b; Iwasaki, *et al.*, 1987; Kaplan *et al.*, 1988) ,and chicken anchorin CII (Sawyer and Cohen, 1985) as represented by the single leter amino acid code are aligned with the inclusion of gaps (-) to emphasize the similarity between the proteins. Residues that are identical to the corresponding residues in endonexin II are indicated by dots (.). The amino terminus of anchorin CII was not unambiguously identified and may extend beyond the sequence shown Fernandez *et al.*, 1988). The location of the highly conserved "consensus sequence" (Funakoshi *et al.*, 1987) that may contain the Ca^{2+} binding site in each of the four repeats is underlined.

Annexins do not have structural similarity with other known Ca^{2+} binding proteins including the "EF-hand" family of proteins (Krefsinger, 1980) and protein kinase C. Although, there is no similarity in primary sequence, it is interesting to note that EF-hand proteins such as calmodulin also arose by duplications of a primordial Ca^{2+}-binding domain.

PHOSPHORYLATION OF THE ANNEXINS

The non-conserved amino terminal domains of the annexins may confer a different biological function to each protein. Some of these functions may be subject to regulation by posttranslational modification because several of the annexins are phosphorylated in this domain. Lipocortin 1 is a high affinity substrate for the epidermal growth factor (EGF) receptor/kinase with phosphorylation occurring on tyrosine-21 (Haigler *et al.*, 1987). The apparent Km for phosphorylation by the EGF receptor/-kinase in isolated A-431 cells is 55 nM (Haigler, 1987) and is the highest known affinity of a tyrosine kinase for a substrate. Lipocortin 1 also is phosphorylated in an EGF-dependent manner in intact cells (Giugni *et al.*, 1985; Sawyer and Cohen, 1985). The stoichiometry of phosphorylation varies as a function of growth state in cultured human fibroblasts. Lipocortin 1 in quiescent cells has less than 0.02 moles phosphate per mole protein while rapidly growing cells that are stimulated with EGF have up to 0.2 moles phosphate per mole protein (Schlaepfer, Giugni, and Haigler, unpublished results). Thus, lipocortin 1 may be a physiological substrate for tyrosine kinases. It is not yet known what effect this phosphorylation has on the biological activity of lipocortin 1. In fact, the biological activity of lipocortin 1 remains a controversial subject as will be discussed in the section on lipocortin as a phospholipase A2 inhibitor.

Lipocortin 2 (also called p36) is phosphorylated by the product of the src oncogene $pp60^{\underline{src}}$ *in vitro* and in intact cells (Erickson *et al.*, 1979; Radke and Martin, 1979). However, there are several lines of evidence that suggest that lipocortin 2 is not a physiological substrate for tyrosine kinases. More experiments are required to define the role of annexins as substrates for tyrosine kinases.

Lipocortins 1 (Schlaepfer and Haigler, 1988), 2 (Gould *et al.*, 1986), and protein II (Weber *et al.*, 1987) are phosphorylated by protein kinase C *in vitro*. Lipocortin 2 (p36) has been shown to be phosphorylated in intact cells but the phosphorylation of the other two proteins in intact cells has not yet been investigated. If these proteins actually are physiological substrates for protein kinase C and for protein-tyrosine kinases, they may play important functions in coordinating the activities of the two separate signaling pathways.

Another important intracellular signaling pathway uses changes in intracellular Ca^{2+} for signal transduction. In the presence of low micromolar Ca^{2+}, annexins bind to phospholipids that are preferentially located on the cytosolic face of the plasma membrane. Thus, changing Ca^{2+} fluxes could cause the proteins to bind to the plasma membrane where the above kinases are located. One annexin, the tetrameric form of lipocortin 2, also associates with the cytoskeleton in a Ca^{2+}-dependent manner (Cheng and Chen, 1981; Gerke and Weber, 1984).

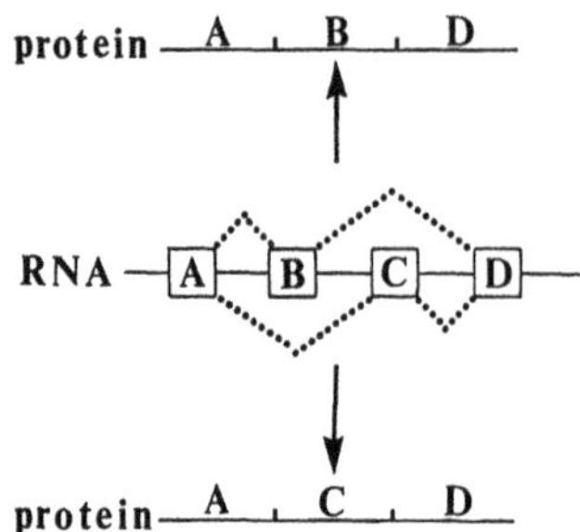

Figure 3. Alternative splicing patterns. Boxes A and D represent constitutively expressed exons (each may be composed of multiple exons) while boxes B and C represent exons that are expressed in a mutually exclusive manner.

COMPARISON OF ENDONEXIN II AND ANCHORIN CII

Endonexin II was first detected as a protein that underwent reversible Ca^{2+}-dependent binding to placental membranes (Haigler *et al.*, 1987; Schlaepfer *et al.*, 1987). It also was identified as an *in vitro* inhibitor of coagulation, "placental anticoagulant protein", (Cheng and Chen, 1981; Funakoshi *et al.*, 1987; Iwasaki *et al.*, 1987) and as an inhibitor of phospholipase A2, "lipocortin V", (Pepinsky *et al.*, 1988). Both inhibitory activities were secondary to Ca^{2+}-dependent binding of phospholipid in the in vitro assays and, thus, may not reflect physiological functions (Haigler *et al.*, 1987; Funakoshi *et al.*, 1987). Although the function of this protein is unknown, the complete amino acid sequence has been deduced from cDNA sequence for both the human (Cheng and Chen, 1981; Funakoshi *et al.*, 1987; Iwasaki *et al.*, 1987; Kaplan *et al.*, 1988; Pepinsky *et al.*, 1988) and rat (Pepinsky *et al.*, 1988) proteins and they share 92% sequence identity (Fig. 2).

Recent structural studies of anchorin CII, a protein thought to act at the cell surface and mediate adhesion to extracellular matrix collagen, have established yet another unexpected relationship with the annexin family. The amino acid sequence of chicken anchorin CII was deduced from its cDNA sequence (Fernandez *et al.*, 1988) and was found to have 40-50% sequence identity with certain annexins (Pfaffle *et al.*, 1988). However, the sequence was not compared to that of endonexin II. As shown in Figure 2, human endonexin II and chicken anchorin CII have striking structural similarity except within a 50 or 52 amino acid segment, respectively, in the second repeat sequence. Outside this segment, the sequence identity is 82%. The similarity is of additional significance because it extends into the amino terminal domain, a region of very limited similarity between other annexins. Based on species variations of other annexins, this is the degree of similarity expected for the same annexin protein in human and chicken. However, the sequence identity abruptly drops to 8% within the 50-52 amino acid segment. In fact, the non-conserved segment is quite acidic in endonexin II and quite basic in anchorin CII. This segment did not share significant sequence similarity with other annexins or proteins in the Gen Bank™ Data Bank.

I propose that human and rat endonexin II and chicken anchorin CII are homologous proteins that arose by alternate splicing of RNA resulting in the incorporation of a 50 amino acid segment into endonexin II and a different 52 amino acid segment into anchorin CII (Figure 3). If this is indeed alternate splicing, it is of an unusual nature. Pairs of exons that are alternately spliced usually represent duplications of ancestral exons and make structurally related protein isoforms (Huang *et al.*, 1986;

Breibart *et al.*, 1987). In contrast, the putative alternatively spliced segments of endonexin II and anchorin CII do not show any sequence similarity.

There are, of course, other possible explanations for the relationship between endonexin II and anchorin CII that do not evoke an alternate splicing mechanism. It also is possible that the two proteins are the products of separate genes that arose by incorporation of different modules into the second half of the second repeat. If two genes exist, will they be present in the same species? If they are, will the two genes exist, will they be present in the same species? If they are, will the two proteins be expressed and serve different biological roles within the same cell? Determination of the structures of these genes will provide important insight into the evolution of these proteins. Clearly, the non-symmetric nature of anchorin CII cannot be explained by the current concept of two sequential duplications of a primitive Ca^{2+}-binding gene.

IMPLICATIONS OF THE RELATIONSHIP BETWEEN ENDONEXIN II AND ANCHORIN CII

Anchorin CII is thought to be located both in the extracellular space and embedded in the extracellular face of chondrocyte membranes and is proposed to mediate cellular adhesion to extracellular matrix collagen (Pfaffle *et al.*, 1988). Although it is tempting to compare the biological characteristics of anchorin CII with those of endonexin II to gain insight into function, there are problems with anchorin CII characterization. The characterizations were performed using an antibody raised against anchorin CII isolated from chondrocyte membranes (Mollenhauer and von der Mark, 1983). However, the amino acid composition and carbohydrate content of isolated anchorin CII are inconsistent with the amino acid composition and glycosylation pattern predicted by the cDNA for anchorin CII (Mollenhauer and von der Mark, 1983; Fernandez *et al.*, 1988). Thus, additional studies using antiserum clearly specific for the protein defined by anchorin CII cDNA are required to define clearly its location and functional properties. Furthermore, anchorin CII cDNA does not predict a signal sequence as would be expected for an extracellular protein. As with certain other annexins, this important point needs closer scrutiny.

LIPOCORTIN 1 AS AN INHIBITOR OF PHOSPHOLIPASE A2

The first complete sequence and the most extensive structural studies of the annexins have been performed at Biogen, Inc. by Drs. Pepinsky and Wallner and their co-workers. They were searching for proteins that were thought to exert an anti-inflammatory affect by inhibition of the enzyme phospholipase A2 (PLA2). A brief history of this area of research is useful in evaluating its current status.

In the early 1980's several groups identified and partially purified proteins that were putative phospholipase A2 inhibitors (DiRosa *et al.*, 1984). They were proposed to mediate the anti-inflammatory effects of steroids by inhibiting PLA2-catalyzed formation of arachidonic acid, the rate-limiting step in prostaglandin, thromboxane, and leukotriene biosynthesis. The proteins were not available in adequate amounts for detailed biochemical analysis such as amino acid sequence determination but, based primarily on immunological comparison, the groups agreed that they were working on identical or closely related proteins. In a joint publication (Wallner *et al.*, 1986) the proteins were named "lipocortin" and assigned the following properties: They are glycoproteins; their synthesis and/or secretion is increased by glucocorticoids; they inhibit PLA2 *in vitro* and *in vivo*; and when phosphorylated, they do not inhibit PLA2. It should be emphasized that these characteristics were not rigorously established. Therefore, this should only be considered a wish list of characteristics that we would like to see in an anti-inflammatory protein.

A major advance occurred when Drs. Pepinsky, Wallner, and co-workers purified significant amounts of a protein from rat peritoneal exudates that inhibited PLA2 *in vitro* (Pepinsky *et al.*, 1986). The human form of this protein, named lipocortin 1, was cloned and expressed in recombinant bacteria (Wallner *et al.*, 1986). Six distinct but structurally-related lipocortins have now been characterized structurally (Pepinsky *et al.*, 1988). In addition to providing a clear structural definition of the proteins, the above studies also facilitate the investigation of their biology by providing milligram quantities of purified proteins, specific antisera, cDNA probes, and recombinant proteins.

A major question is whether the putative PLA2 inhibitors that were biologically defined in the early 1980's (DiRosa *et al.*, 1984) and named "lipocortins" are the same proteins as the six proteins, lipocortins 1 through 6, that were isolated and structurally characterized by Pepinsky, Wallner and their colleagues at Biogen (Wallner *et al.*, 1986; Pepinsky *et al.*, 1988). This question must begin with a detailed consideration of the *in vitro* assay used by the Biogen group to identify the proteins. The assay measured inhibition of pancreatic PLA2-catalyzed release of radioactivity from [3H]oleic acid-labeled *E. coli* membranes (Pepinsky *et al.*, 1986). Independent studies have shown that PLA2 inhibition in this assay was secondary to Ca^{2+}-dependent binding of the annexin to the lipid substrate and not by way of direct interaction with the enzyme (Davidson *et al.*, 1987; Haigler *et al.*, 1987). This "substrate depletion" mechanism suggests that this assay does not measure a physiologically relevant inhibition. It is generally agreed that the *in vitro* PLA2 assays measured Ca^{2+}-dependent binding of protein to lipid and that the assays were not reliable measures of the protein's physiological role. To date, there is no direct evidence for high affinity interactions of annexins with PLA2.

Little has been published concerning the biological properties of lipocortins 1 through 6 in intact cells. In fact, the basic question of whether the proteins are found extracellularly has not been clearly resolved. Although lipocortin 1 originally was purified from peritoneal exudates (Pepinsky *et al.*, 1986), subsequent studies have found lipocortin 1 to be an intracellular protein. All other annexins that have been studied also are found intracellularly. A secretory signal sequence is not found within the mRNA's of these proteins so additional studies are required to directly show whether the proteins are secreted and, if so, to determine the mechanism by which secretion occurs. If lipocortins do play a physiological role at an extracellular location, it is important to identify their target cells and characterize the interactions.

Recombinant lipocortin 1 has been reported to inhibit thromboxane release from guinea pig perfused lung (Cirino *et al.*, 1987). It is important to determine whether the effects of the exogenously added protein are physiological or pharmacological. The important question of whether synthesis and/or secretion of lipocortins 1 through 6 are stimulated by glucocorticoids has not been answered clearly. Published reports have been negative on this point (Wallner *et al.*, 1986).

FUTURE DIRECTIONS FOR RESEARCH IN ANNEXINS

When it was first shown that the putative PLA2 inhibitors lipocortin 1 and 2 were identical to the cellular substrates for tyrosine-protein kinases, p35 and p36, there was optimism that these studies would rapidly unlock the secrets of both the steroid-induced anti-inflammatory response and the mitogenic pathway. However, studies of the biological roles of these proteins are moving slowly compared to the structural studies. Now it is not even generally accepted that lipocortin 1 through 6 are actually physiological inhibitors of PLA2. It is important to determine if these structurally-defined proteins possess the same biological properties originally attributed to "lipocortins".

The primary sequence of certain of the proteins in this family have been determined in several species and found to be highly conserved; e.g. the sequences of murine, bovine, and human lipocortin 2 differ by only 2%. For a protein to be this highly conserved, there must be constraints on sequence drift over much of the polypeptide chain. This may indicate that the annexins interact at multiple sites with other molecules and that each site exerts selection pressure. The proteins are known to interact with phospholipid, Ca^{2+}, and several proteins. The interactions with other proteins have been proposed and, considering the history of this exciting area of research, other surprises may be forthcoming.

Since several of the annexins contain conserved sites for phosphorylation by protein-tyrosine kinases and protein kinase C, one of the most intriguing possible roles is in intracellular signal transduction. Additional experiments are required to determine the biological activity of these proteins and to determine if phosphorylation is a physiological regulator of this activity.

REFERENCES

Breitbart R.E., Andreadis A., and Nadal-Ginard B. 1987. Alternative splicing:a ubiquitous mechanism for the generation of multiple protein isoforms from single genes. Ann. Rev. Biochem. 56:467.

Burns A.L., Magendzo K., Shirvan A., Srivastava M., Rojas E., Alijani M.R., and Pollard H. 1989. Calcium channel activity of purified human synexin and structure of the synexin gene. Proc. Natl. Acad. Sci. (USA) in press.

Cheng Y.S-E. and Chen L.B. 1981. Detection of phosphotyrosine-containing 36,000 dalton protein in the framework of cells transformed with Rous sarcoma virus. J. Cell Biol. 104:503.

Cirino G., Flower R.J., Browning J.L., Sinclair L.K., and Pepinsky R.B. 1987. Recombinant human lipocortin I inhibits thromboxane release from guinea pig isolated perfused lung. Nature (London) 328:270.

Creutz C.E., Zaks W.J., Hamman H.C., Crane S., Martin W.H., Gould K.L., Oddie K.M., and Parsons S.J. 1987. Identification of chromaffin granule-binding proteins. J. Biol. Chem. 262:1860.

Crompton M.R., Moss S.E., and Crompton M.J. 1988. Diversity on the lipocortin/calpactin family. Cell 55:1.

Crompton M.R., Owens R.J., Totty N.F., Moss S.W., Waterfield M.D., and Crompton M.J. 1988. Primary structure of the human, membrane-associated Ca^{2+}-binding protein p68: a novel member of a protein family. EMBO J. 7:21.

Davidson F.F., Dennis E.A., Powell M., and Glenney J.R. 1987. Inhibition of phospholipase A2 by "lipocortins" and calpactins. J. Biol. Chem. 262:1698.

De B.K., Misono K.S., Lukas T., Mroczkowski B., and Cohen S. 1986. A calcium-dependent 35 kilodalton substarte for epidermal growth factor receptor/kinase isolated from normal tissue. J. Biol. Chem. 261:13784.

DiRosa M., Flower R., Hirata F., Parente L., and Russo-Marie F. 1984. Anti-phospholipase proteins. Prostaglandins 28:441.

Erikson R.L., Collett M.S., Erikson E., Purchio A.F., and Brugge J.S. 1979. Protein phosphorylation mediated by partially purified avian sarcoma virus transforming gene product. Cold Spring Harbor Symp. Quant. Biol. 44:907.

Fernandez M.P., Selmin O., Martin G.R., Yamada Y., Pfaffle M., Deutzmann R., Mollenhauer J. and von der Mark K. 1988. The structure of anchorin II, a collagen binding protein isolated from chondrocyte membrane. J. Biol. Chem. 263:5921.

Funakoshi T., Heimark R.L., Hendrickson L.E., McMullen B.A. and Fujikawa K. 1987a. Human placental anticoagulant protein: Isolation and characterization. Biochem. 26:5572.

Funakoshi T., Hendrickson L.E., McMullen B.A. and Fujikawa K. 1987b. Primary structure of human placental anti-coagulant protein. Biochem. 26:8087.

Geisow M.J., Fritsche U., Hexham J.M., Dash B. and Johnson T. 1986. A consensus amino acid sequence repeat in *Torpedo* and mammalian Ca^{2+}-dependent membrane binding proteins. Nature (London) 320:636.

Gerke V. and Weber K. 1984. Identity of a p36k phosphorylated upon Rous sarcoma virus transformation with a protein purified from brush borders: calcium dependent binding to non-erythroid spectrin and F-actin. EMBO J. 3:227.

Giugni T.D., James L.C., and Haigler H.T. 1985. Epidermal growth factor stimulates tyrosine phosphorylation of specific proteins in permeabilized human fibroblasts. J. Biol. Chem. 261:15081.

Glenney J.R. 1986. Phospholipid dependent Ca^{2+}-binding by the 36kd tyrosine kinase substrate (calpactin) and its 33kd core. J. Biol.Chem. 261:7247.

Glenney J.R. 1986. Two related but distinct forms of the 36000 M tyrosine substrate (calpactin) which interact with phospholipid and actin in a Ca^{2+}-dependent manner. Proc. Natl. Acad. Sci. (USA) 83:4258.

Glenney J.R., Jr. and Tack B. 1985. Amino-terminal sequence of p36 and associated p10: identification of the site of tyrosine phosphorylation and homology with S-100. Proc. Natl. Acad. Sci. (USA) 82:7884.

Gould K.L., Woodgett J.R., Isacke C.M., and Hunter T. 1986. The protein-tyrosine kinase substrate p36 is also a substrate for protein kinase C *in vitro* and *in vivo*. Mol. Cell. Biol. 6:2738.

Haigler H.T., Schlaepfer D.D., and Burgess W.H. 1987. Characterization of lipocortin I and an immunologically unrelated 33kDa protein as epidermal growth factor receptor/kinase substrates and phospholipase A2 inhibitors. J. Biol. Chem. 262:6921.

Huang K.-S., Wallner B.P., Mattaliano R.J., Tizard R., Burne C., Frey A., Hession C., McGray P., Sinclair L.K., Chow E.P., Browning J.L., Ramachandran K.L., Tang J., Smart J.E. and Pepinsky R.B. 1986. Two human 35kd inhibitors of phospholipase A2 are related to substrate of pp60 $^{v\text{-}src}$ and of the epidermal growth factor receptor / kinase. Cell 46:191.

Iwasaki A., Suda M., Nakao H., Nagoya T., Saino Y., Arai K., Mizoguchi T., Sato F., Yoshizaki H., Hirata M., Miyata T., Shidara Y., Murata M., and Maki M. 1987. Structure and expression of cDNA for an inhibitor of blood coagulation isolated from human placenta: A new lipocortin-like protein. J. Biochem. 102:1261.

Iwata M. and Ishizaka K. 1987. *In vitro* modulation of antigen-primed T cells by a glycosylation-inhibiting factor that regulates the formation of antigen-specific suppressive factors. Proc. Natl. Acad. Sci. (USA) 84:2444.

Kaplan R., Jaye M., Burgess W.H., Schlaepfer D.D., and Haigler H.T. 1988. Cloning and expression of cDNA for human endonexin II, a Ca^{++} and phospholipid binding protein. J. Biol. Chem. 263:8034.

Kretsinger R.H. 1980. Structure and evolution of calcium-modulated proteins. CRC Crit. Rev. Biochem. 8:119.

Mollenhauer J. and von der Mark K. 1983. Isolation and characterization of a collagen-binding glycoprotein from chondrocyte membranes. EMBO J. 2:45.

Pepinsky R.B., Sinclair L.K., Browning J.L., Mattaliano R.J., Smart J.E., Chow E.P., Falbel T., Ribolini A., Garwin J.L., and Wallner B.P. 1986. Purification and partial sequence analysis of a 37-kDa protein that inhibits phospholipase A2 activity from rat peritoneal exudates. J. Biol. Chem. 261:4239.

Pepinsky R.B., Tizard R., Mattaliano R.J., Sinclair L.K., Miller G.T., Browning J.L., Chow E.P., Burne C., Huang K.-S., Pratt D., Wachter L., Hession C., Frey A.Z., and Wallner B.P. 1988. Five distinct calcium and phospholipid binding proteins share homology with lipocortin I. J. Biol. Chem. 263:10799.

Pfaffle M., Roggiero R., Hofmann H., Fernandez M.P., Selmin O., Yamada Y., Garrone R., and von der Mark K. 1988. Biosynthesis, secretion and extracellular localization of anchorin II, a collagen-binding protein of the calpactin family. EMBO J. 7:2335.

Radke K. and Martin G.S. 1979. Transformation by Rous sarcoma virus: Effects of src gene expression on the synthesis and phosphorylation of cellular polypeptides. Proc. Natl. Acad. Sci. (USA) 70:5212.

Saris C.J.M., Tack B.F., Kristensen T., Glenney J.R. and Hunter T. 1986. The cDNA sequence for the protein-tyrosine kinase substrate (calpactin I heavy chain) reveals a multidomain protein with internal repeats. Cell 4:201.

Sawyer S.T. and Cohen S. 1985. Epidermal growth factor stimulates the phosphorylation of the calcium-dependent 35,000-dalton substrate in intact A-431 cells. J. Biol. Chem. 260:8233.

Schlaepfer D.D., Mehlman T., Burgess W.H., and Haigler H.T. 1987. Characterization of Ca^{2+}-dependent phospholipid binding and phosphorylation of Lipocortin I. Proc. Natl. Acad. Sci. (USA) 84:6078.

Schlaepfer D.D. and Haigler H.T. 1987. *In vitro* protein kinase C phosphorylation sites of placental lipocortin. J. Biol Chem. 262:6931.

Schlaepfer D.D. and Haigler H.T. 1988. Structural and functional characterization of endonexin II, a calcium-and phospholipid-binding protein. Biochem. 27:4253.

Shadle P.J., Gerke V., and Weber K. 1985. Three Ca^{++}-binding proteins from porcine liver and intestine differ immunologically and physicochemicall and are distinct in Ca^{++} affinities. J. Biol. Chem. 260:16354.

Smith V.L. and Dedman J.R. 1986. An immunologic comparison of several novel calcium-binding proteins. J. Biol. Chem. 261:15815.

Sudhof T.C., Slaughter C.A., Leznicki I., Barjon P., and Reynolds G.A. 1988. Human 67-kDa calelectrin contains a duplication of four repeats found in 35kDa lipocortins. Proc. Natl. Acad. Sci. (USA) 85:664.

Wallner B.P., Mattaliano R.J., Hession C., Cate R.L., Tizard R., Sinclair L.K., Foeller C., Chow E.P., Browning J.L., Ramachandran K.L. and Pepinsky R.B. 1986. Cloning and expression of human lipocortin; a phospholipase A2 inhibitor with potential antiinflammatory actvity. Nature (London) 320:77.

Weber K., Johnsson N., Plessmann U., Van P.N., Soling H.-D., Ampe C., and Vandekerckhove J. 1987. The amino acid sequence of protein II and its phosphorylation site for protein kinase C; the domain structure of Ca^{++}-modulated lipid-binding proteins. EMBO J. 6:1599.

CONSIDERATION OF THE ROLE OF THE CELL CYCLE IN GROWTH FACTOR MODULATED RESPONSES

W. J. Pledger and
C. J. Morgan

Department of Cell Biology
Vanderbilt University School of Medicine
Nashville, TN 37232

The interval between consecutive mitoses has been termed the cell cycle and during this interval the cell must increase cellular mass by duplicating cellular components, replicate its DNA, and undergoing an orderly division to equal progeny cells (Baserga, 1984, 1985). The modulation of G_1 events by growth factors provide much of the regulation for cellular proliferation. Normal cells become growth-arrested *in vitro* when they reach confluent density in culture conditions; cells may also be arrested at subconfluent densities by depriving them of serum growth factors. These quiescent cells are reversibly growth arrested in the G_0 phase. When G_0-arrested fibroblasts are stimulated to re-enter the cell cycle by the addition of fresh serum, there is a characteristic lag before the onset of DNA synthesis. This lag describes the G_1 phase of the cell cycle. A fibroblastic culture system provides a unique way to investigate the growth factor regulation of G_1-specific events.

Serum can be fractionated into two classes of components which regulate two different aspects of cell function, the initiation of cellular proliferation and the regulation of G_0/G_1 traverse (Pledger *et al.*, 1987). Platelet-derived growth factor (PDGF) released from platelets into serum during the clotting process is the serum factor that stimulates the initiation of the cell cycle in fibroblasts. This initiation of the growth response has been termed competence formation. The factors required for the traverse of the G_1 phase are present in the plasma portion of blood and are termed progression factors. The identification of these two classes of serum factors provide the experimental basis for a model of cell cycle regulation in which the process of commitment to cell division can been separated into two discrete stages regulated by two different classes of serum growth factors.

COMPETENCE FORMATION

PDGF is a heat-stable cationic protein of approximately 30,000 daltons. The protein core of PDGF is disulfide link dimer composed of two distinct homologous polypeptides, A and B, which are encoded by different genes (Kelly *et al.*, 1985; Sejersen *et al.*, 1986). The PDGF can exist *in vivo* as a homodimer of either A chains or B chains and as a heterodimer. The B chain of PDGF is the gene product of the proto-oncogene c-sis -- the cellular homologue of the v-sis transforming gene of the simian sarcoma retrovirus (Doolittle *et al.*, 1986). The initial interaction of PDGF with its target cell is by non- covalent binding to a cell surface receptor with high affinity and selectivity for PDGF. The PDGF receptor is a transmembrane glycoprotein of approximately 180,000 daltons. The sequence of the cDNA encoding the entire murine PDGF receptor indicates that the protein is synthesized with a 32 amino acid signal peptide which is cleaved to produce a 1,066 amino acid mature protein (Yarden *et al.*, 1986). The individual types of PDGF receptor present on the cell is at present unclear. It has been suggested that there are at least two different PDGF receptors which recognize different homo- or heterodimer PDGF molecules. At present, conflicting data make it difficult to determine the exact nature of the PDGF receptors.

The binding of PDGF to its receptor sets in motion a series of events which leads to competence formation. The state of competence in the cell is more stable than the association of the PDGF molecule with its receptor (Singh *et al.*, 1983). That is, the PDGF receptor complex has a half life of approximately one to two hours, whereas the competence state has a half life of approximately fifteen to eighteen hours. A competent cell differs from a quiescent cell in that the competent cell can respond to growth factors present in plasma and progress through the G_1 phase and enter into DNA synthesis. It is believed that genes induced by PDGF render the cell competent. The exact genes and the function of their gene products required for competence formation is unknown. The induction of the c-myc gene by PDGF has been suggested to be responsible for competence formation.

Binding of PDGF to its cell surface receptor is followed within minutes by rapid receptor autophosphorylation and the phosphorylation of tyrosine on a number of endogenous proteins may be via the PDGF receptor/tyrosine kinase. The cytoplasmic portion of the PDGF receptor has been shown to contain a tyrosine kinase activity which not only catalyzes the receptor autophosphorylation but can also bring about the phosphorylation of other substrates. PDGF activates the phosphatidylinositol-specific phospholipase C which hydrolyses plasma phosphatidylinositol diphosphate to form inositol triphosphate and diacylglycerol (Habenicht *et al.*, 1981; Berridge, 1987). Inositol-triphosphate can cause the release of calcium ion from an intercellular pool. This results in an increase in cytoplasmic calcium ion concentration. The diacylglycerol generated in this manner is an activator of the serine/threonine specific protein kinase C which is capable of phosphorylating a number of intercellular proteins (Rassmussen *et al.*, 1984; Downward *et al.*, 1985).

It is not known if the change in calcium concentration or the activation of protein kinase C is directly involved in inducing the genes which render the cell competent. TPA, an activator of protein kinase C can stimulate a proliferative response and induce the expression of genes associated with competence. Table 1 shows many cellular responses that are modulated by the addition of PDGF to quiescent cells. The involvement of these rapid PDGF-induced cellular responses in competence formation are not known.

These initial early events have been shown to be followed by the synthesis of new gene products. PDGF has been shown to stimulate rapidly the synthesis of several unique proteins. One of these proteins, CpI, was shown to be associated with the nuclear portion of the cell. Other genes have been shown to be induced by PDGF including c-fos, c-myc, jun-

Table 1. Rapid Cellular Responses to PDGF

1. Activation of the PDGF receptor kinase activity.
2. Stimulation of phospholipase C activity.
3. The release of vinculin from adhesion plaques (Herman & Pledger, 1985).
4. The dissolution of actin stress cables (Herman & Pledger, 1985).
5. Increased intercellular calcium.
6. Alkalization of the cell (L'Allemain *et al.*, 1984).
7. Stimulation of synthesis of progression growth factor receptors (Clemmons *et al.*, 1980).

R, and KC (this is an abreviated list). The gene products of c-fos and c-myc have been shown to be nuclear proteins and c-fos has been shown to participate in control of transcription. The unscheduled production of PDGF, c-fos or c-myc has been shown to alter normal growth control and function as oncogenes.

PROGRESSION

The characteristic G_1 time for quiescent fibroblasts stimulated to undergo cellular proliferation is twelve to fifteen hours. This delay is longer than the G_1 period in continuously cycling cells; experimentally growing murine BALB/c-3T3 fibroblasts proliferate with a six hour G_1 phase compared with the twelve hours required for density-arrested BALB/c-3T3 cells to enter S phase after serum stimulation. The growth factors present in plasma allow competent cells to initiate DNA synthesis after a twelve hour lag; this lag time represents the minimal time required for density-arrested cells to transit the G_1 phase after stimulation of the cell cycle. The transient exposure of competent cells to plasma have identified two arrest points within the G_1 phase that were termed the "V" point and "W" point (Pledger *et al.*, 1978). The "V" point occurs six hours before the onset of DNA synthesis, bisecting the G_1 phase, and the "W" point is immediate prior to DNA synthesis and appears to be the point of commitment to DNA synthesis and cell division. Epidermal growth factor (EGF) and insulin-like growth factor (IGF-I) can replace the plasma requirements with progression through G_1 (Leof *et al.*, 1981). IGF-I has been shown to be one of the components in plasma required for progression (Russell *et al.*, 1984). As illustrated below, EGF and IGF-I are required for competent cells to reach the "V" point but only IGF-I is required for transit from the "V" point to the G_1/S boundary and commitment to DNA synthesis (Leof *et al.*, 1983).

The expression of IGF-I receptors in fibroblasts is influenced by PDGF. Transient exposure of growth-arrested cells to PDGF followed by platelet-poor plasma caused a two-fold increase in IGF-I binding (Clemmons *et al.*, 1980). This increase in IGF-I binding reflected a doubling in the receptor numbers on the cell. The function of these two growth factors (EGF and IGF-I) in the progression of competent cells is

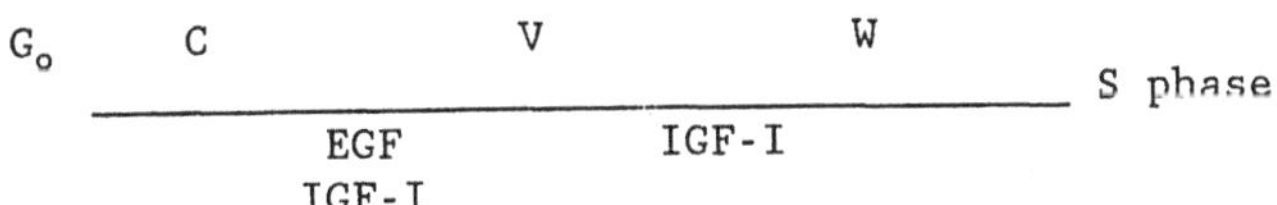

Figure 1. Growth factor requirements during G_1 phase for competent BALB/c-3T3 fibroblasts to enter s phase.

not known. Treatment of cells in early or late G_1 with inhibitors of protein and RNA synthesis have shown that, although protein synthesis is required throughout most of the G_1 period, the traverse of cells through late G_1 is relatively refractory to inhibition of RNA synthesis (Olashaw *et al.*, 1987). Moreover, progression from the "V" point to S phase does not appear to require increased overall protein synthesis, nor does IGF-I appear to stimulate synthesis of specific proteins detected by two dimensional electrophoresis. Experiments have shown that IGF-I brings about post-translational modification of several proteins. It is not known if these modifications are responsible for the IGF-I permissive action in allowing progression through late G_1.

The cellular events bringing about commitment to DNA synthesis at the G_1/S phase boundary is not known. This biochemical process is the final regulatory event leading to the commitment of the cell cycle and cellular proliferation.

COORDINATE CONTROL OF STIMULATION-INHIBITION OF CELLULAR PROLIFERATION

Six to eight hours after exposure to PDGF, BALB/c-3T3 fibroblasts produce mRNA encoding interferon and mRNA levels remain high for as long as twelve hours after stimulation (Zullo *et al.*, 1985). Addition of interferon to arrested cells stimulated with PDGF and plasma results in inhibition only if interferon is added in the early part of the G_1 phase suggesting that interferon inhibits events involved in the early transitions associated with induction of competence. Consistent with this idea is the observation that interferon inhibits the overall stimulation of protein synthesis by PDGF and suppresses the induction of the competence related proteins such as CpI (Lin *et al.*, 1986). Interferon has been shown to inhibit other early gene responses to PDGF such as the response of c-myc. The involvement of the inhibition of cellular proliferation by interferon *in vivo* is not known, but other potent inhibitors of cell proliferation are known for other cell types. For example, TGFβ is a potent inhibitor of epithelial cell proliferation.

CELL CYCLE DEPENDENT GROWTH FACTOR STIMULATION OF GENE EXPRESSION

The twelve hour G_1 phase is subdivided by not only plasma dependent arrest points but nutritional arrest points. For example, amino acid starvation, high intracellular levels of cAMP, and depletion of IGF-I brings about growth arrest at the "V" point (O'Keefe & Pledger, 1983). In addition, cells can be synchronized in S phase by a variety of reversible inhibitors of DNA synthesis such as hydroxyurea, aphidicolin and methotrexate. These arrest points can be used to design protocols to compare the regulation of gene expression during different phases of the cell cycle. When G_0 cells are stimulated by PDGF, c-fos and c-myc are rapidly induced. However, when G_0 cells are treated with plasma, there is no increase in the synthesis of the c-fos or c-myc mRNA. We have used methotrexate and aphidicolin to inhibit cells in early S phase. When cells are inhibited in early S phase and treated with plasma, there is no induction of fos or myc. However, if the cells that were inhibited in early S phase are treated with PDGF, there is a rapid transient increase in fos and myc RNA synthesis. After this induction by PDGF, the cells can respond to plasma and produce a continued increase in the synthesis of c-myc mRNA. Conversely, in contrast to these results, plasma does not induce c-fos message. The induction of the c-myc message in S phase by plasma is not a stabilization of message as was shown with actinomycin D stability experiments.

In general, the model system described can be used to study the effects of the cell cycle in regulating the response of cells to various growth factors in addition to the study of the mechanism of action of growth factors at the point in the cell cycle where that growth factor normally functions. Information gained on the normal regulation of cellular proliferation by growth factors will help us understand how normal growth control is abrogated to bring about the development of neoplastic growth.

REFERENCES

Baserga R. 1985. The Biology of Cell Reproduction. Harvard University Press. Cambridge, Mass.

Baserga R. 1985. Growth in size and cell DNA replication. Exp. Cell Res. 151:1.

Berridge M. 1987. Inositol phosphate and diacylglycerol: Two interacting second messengers. Ann. Rev. Biochemistry 56:159.

Clemmons D.R., Van Wyk J.J., Pledger W.J. 1980. Sequential addition of platelet factor and plasma to BALB/c-3T3 fibroblast cultures stimulates somatomedin C binding early in cell cycle. Proc. Natl. Acad. Sci. (USA) 77:6644.

Doolittle R.F., Hunkapiller M.W., Hood L.E., Devare S.G., Robbins K.C., Aaronson S.A., and Antoniades H.N. 1983. Simian sarcoma virus onc gene, v-sis, is derived from the gene (or genes) encoding a platelet-derived growth factor. Science 221:275.

Downward J., Waterfield M.D., and Parker P.J. 1985. Autophosphorylation and protein kinase C phosphorylation of the epidermal growth factor receptor: Effect on tyrosine kinase activity and ligand binding affinity. J. Biol. Chem 260:14538.

Habenicht A.J.R., Glomset J.A., King W.C., Nist C., Mitchell C.D., and Ross R. 1981. Early changes in phosphatidylinositol and arachidonic acid metabolism in quiescent Swiss 3T3 cells stimulated to divide by platelet-derived growth factor. J. Biol. Chem. 256:12329.

Herman B. and Pledger W.J. 1985. Platelet-derived growth factor-induced alterations in vinculin and actin distribution in BALB/c-3T3 cells. J. Cell Biol. 100:1031.

Kelly J.D., Raines E.W., Ross R., and Murray M.J. 1985. The B chain of PDGF alone is sufficient for mitogenesis. EMBO J. 4:3399.

L'Allemain G., Paris S., and Pouyssegur J. 1984. Growth factor activation and intracellular pH regulation in fibroblasts: Evidence of a major role of the Na+/H+ antiprot. J. Biol. Chem. 259:5809.

Leof E.B., Wharton W., Van Wyk J.J., and Pledger W.J. 1981. Epidermal growth Factor (EGF) and somatomedin C regulate G_1 progression in competent BALB/c-3T3 cells. Exp. Cell Res. 141:107.

Leof E.B., Van Wyk J. J., O'Keefe E.J., and Pledger W.J. 1983. Epidermal growth factor (EGF) is required only during the traverse of early G_1 in PDGF stimulated density-arrested BALB/c-3T3 cells. Exp. Cell Res. 147:202.

Lin S.L., Kikuchi T., Pledger W.J., and Tamm I. 1986. Interferon inhibits the establishment of competence in G_0/S phase transition. Science 233:356.

O'Keefe E.J. and Pledger W.J. 1983. A model of cell cycle control. Mol. Cell. Endocrinol. 31:167.

Olashaw N.E., Van Wyk J.J., and Pledger W.J. 1987. Control of late G_0/G_1 progression and protein modification by somatomedin C/insulin-like growth factor-1. Am. J. Physiol. Cell Physiol. 22:575.

Pledger W.J., Stiles C.D., Antoniades H. N., and Sher C.D. 1977. Induction of DNA synthesis in BALB/c-3T3 cells by serum components. Reevaluation of commitment process. Proc. Natl. Acad. Sci. (USA) 74:4481.

Pledger W.J., Stiles C.D., Antoniades H.N., and Scher C.D. 1978. An ordered sequence of events is required before BALB/c-3T3 cells become committed to DNA synthesis. Proc. Natl. Acad. Sci. (USA) 75:2839.

Rassmussen H., Kojima I., Kojima K., Zawalich W., and Apfeldorf W. 1984. Calcium as intracellular messenger: Sensitivity modulation, C-kinase pathway, and sustained cellular response. Adv. Cyclic Nuc. Res. Prot. Phosphor. 18:159.

Russell W.E., Van Wyk J.J., and Pledger W.J. 1984. Inhibition of the mitogenic effects of plasma by a monoclonal antibody to somatomedin C. Proc. Natl. Acad. Sci. (USA) 81:2389.

Sejersen T., Betsholtz C., Sjolund M., Heldin C.-H., Westermakr B., and Thyberg J. 1986. Rat skeletal myoblasts and arterial smooth muscle cells express the gene for the A chain but not the gene for the B chain (c-sis) of platelet-derived growth factor (PDGF) and produce a PDGF-like protein. Proc. Natl. Acad. Sci. (USA) 83:6844.

Singh J.P., Charkin M.A., Pledger W.J., Scher C.D., and Stiles C.D. 1983. Persistence of competence formation. J. Cell. Biol. 96:1497.

Yarden Y., Escobedo J.A., Kung W.-J., Yang-Feng T.L., Daniel T.O., Tremble P.M., Chen E.Y., Ando M.E., Harkins R.N., Franke U., Fried V.A., Ullrich A., and Williams L.T. 1986. Structure of the receptor for platelet-derived growth factor helps define a family of closely related growth factor receptors. Nature (London) 323:226.

Zullo J.N., Cochran B.H., Huang A.S., and Stiles C.D. 1985. Platelet-derived growth factor and double-stranded ribonucleic acids stimulate expression of the same genes in 3T3 cells. Cell 43:793.

ONCOGENIC TRANSFORMATION BY BASIC FIBROBLAST GROWTH FACTOR

Snezna Rogelj and
Michael Klagsbrun

Children's Hospital
Harvard Medical School
320 Longwood Avenue
Boston, MA 02115

Several years ago Sporn and Todaro proposed autocrine growth stimulation as a mechanism for cellular transformation (Sporn and Todaro, 1980). According to this model, continuous proliferation occurs in a cell that possesses the cognate receptor for a growth factor that the cell secretes, (Long *et al.*, 1985; Rosenthal *et al.*, 1986; Finzi *et al.*, 1987; Watanabe *et al.*, 1987). This has been shown to be the cause of oncogenic conversion by numerous growth factors, including sis, epidermal growth factor (EGF), CSF-1, GM-CSF, and transforming growth factor α (TGF α) (Doolittle *et al.*, 1983; Rosenthal *et al.*, 1986; Stern *et al.*, 1987; Waterfield *et al.*, 1983; Wheeler *et al.*, 1986).

Where within the cell does the mitotically productive interaction of growth factor-receptor occur? It is possible that receptor-ligand interaction occurs at the cell suface. In this case, the transforming growth factor must first be secreted into the extracellular space. Alternatively, it is possible that the receptor-ligand interaction occurs earlier, in the endoplasmic reticulum of the secretory pathway. The receptor-ligand complex may activate the mitotic pathway at the cell surface, or at an earlier site (Leal *et al.*, 1985; Keating and Williams, 1987; Robbins *et al.*, 1988).

Basic fibroblast growth factor (bFGF) is a potent mitogen for cells of mesodermal and neuroectodemal origin. It is synthesized by a variety of tissues and cell lines. In all cases, it remains associated with the cells rather than being secreted into the extracellular medium. This cellular association may be due to the lack of the signal peptide domain in the bFGF protein. For example, endothelial cells *in vivo* and *in vitro* synthesize intracellular bFGF, as well as respond to this growth factor when it is added exogenously (Schweigerer, 1987). However, bFGF produced by the endothelial cells themselves does not induce their continuous proliferation. It is possible that bFGF produced by a cell is localized to a different subcellular compartment, resulting in a physical separation from the receptor, thus avoiding the auto-stimulatory circuit.

As such, bFGF promised to be a particulary useful reagent for testing the model of autocrine growth, and determining the site of stimulation in angiogenesis. Utilizing this natural character of bFGF structure we asked the following questions pertaining to the mechanism of autocrine

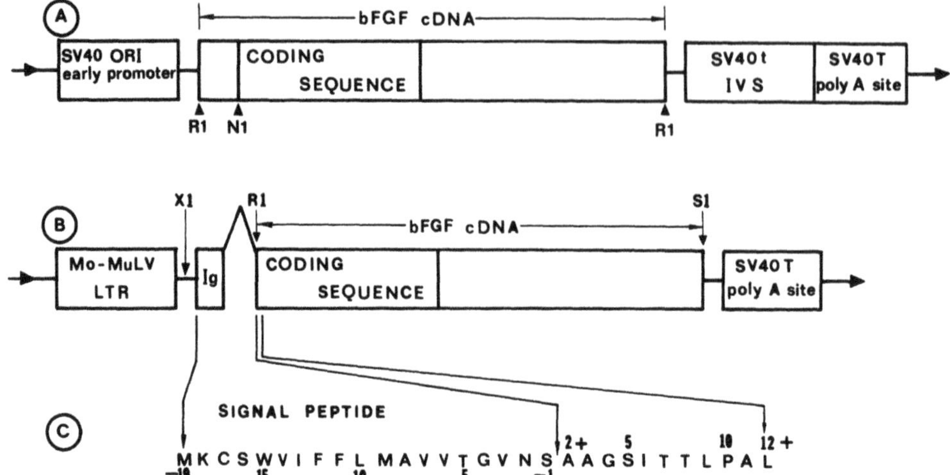

Figure 1. Basic FGF expression constructs (R1 = Eco R1, X1 = Xbal, S1 = Sall, N1 = Ncol restriction site). A. pbFGF plasmid: constitutuive expression of the native bovine brain bFGF is driven by the SV40 early promoter. B. pIgbFGF plasmid: constitutive expression of the immunoglobulin signal peptide - bFGF chimeric protein is driven by the murine leukemia virus long terminal repeat (LTR). C. The amino terminus of the predicted primary translation product of the signal peptide-bFGF fusion protein where the 19 amino acid signal peptide is fused to the second amino acid of the native bovine brain bFGF protein.

transformation: Do bFGF responsive cells induced to constitutively produce high amounts of non-secreted native bFGF respond by becoming transformed, or are such cells required to produce a secreted form of bFGF to undergo autocrine transformation?

In order to address this question we constructed two mammalian expression vectors: one that induces expression of the native, (signal peptide lacking) bFGF molecule, and one that induces expression of a secretable form of bFGF. For expression of the native bFGF, the bovine brain bFGF cDNA (Abraham *et al.*, 1986) was inserted into pJay3 expression vector (Figure 1A; Rogelj *et al.*, 1988). In the second clone the sequence encoding the amino terminal immunoglobulin signal peptide of 19 amino acids was fused to the second amino acid of cDNA encoded bFGF molecule (Figure 1B; Loh *et al.*, 1983; Stern *et al.*, 1987). The predicted primary translation product of the chimeric signal peptide-bFGF (sp-bFGF) is illustrated in Figure 1C. This chimeric protein should be inserted into endoplasmic reticulum, the signal peptide cleaved off, yielding the processed protein almost identical in size to the native 18 Kd bFGF.

Clonal cell lines expressing the native and the chimeric bFGF proteins were derived by co-transfécting these expressin constructs with dominant selectable marker pSV2-neo (Southern and Berg, 1982) into NIH-3T3 fibroblasts that express bFGF receptors. Production of bFGF by these cell lines was demonstrated by first showing that the lysates contain cell-associted heparin binding growth factor activity eluting with the characteristic 1.5M NaCl (Klagsbrun *et al.*, 1986). Further confirmation that the growth activity was due to bFGF was provided by Western blot analysis using antibodies raised against an internal region of bFGF (Klagsbrun *et al.*, 1986).

For further studies three cell lines were selected: NIH-N cell line, expressing the neomycin resistance gene alone; NIH-BNM7 cell line expressing the native bFGF molecule; and NIH-IgBNM6-1 cell line expressing the chimeric sp-bFGF protein. The two bFGF expressing cell lines were chosen because they express 0.6-0.7 units of bFGF activity. This is about as much as is produced by bovine aortic endothelial cells, and about 20 times as much as is produced by the control NIH-N cell line.

Production of bFGF by these cell lines is illustrated in Figure 2. Cell lysates and conditioned medium from equal numbers of NIH-N, NIH-BNM7 and NIH-IgBNM6-1 cell lines were analyzed by heparin affinity chromatography. As shown in Figure 2A, both NIH-BNM7 and NIH-IgBNM6-1 cell lines elaborate growth factor activity that elutes at the characteristic 1.5 M NaCl. No bFGF activity elutes from the heparin-sepharose column loaded with the control NIH-N cell lysate. The fractions eluting with 1.5 M NaCl were probed in an immunoblot using an antibody raised against an internal region of bFGF (Figure 2B; Klagsbrun *et al.*,1986 Vlodavsky *et al.*, 1987). Both the NIH-BNM7 and NIH-IgBNM6-1 cells express an 18 Kd immunoreactive species that comigrates with the purified recombinant bovine brain bFGF used as a positive control. The presence of this species in the NIH-IgBNM6-1 cell line suggests that the chimeric sp-bFGF protein has entered the endoplasmic reticulum where the signal peptide was cleaved off. In addition to this processed protein, immunoreactive proteins of about 20, 24, 30 and 34 Kd are present in the NIH-IgBNM6-1 cell line. These higher molecular weight proteins may represent the uncleaved form of sp-bFGF, as well as bFGF molecules post-translationally modified in the endoplasmic reticulum.

In contrast to the considerable amount of bFGF activity found associated with the cells, no such activity was found in the conditioned medium (Figure 2A). The lack of secreted bFGF in the NIH-IgBNM6-1 cell line elaborating the chimeric sp-bFGF protein was not due to rapid degradation since addition of heparin as bFGF stabilizing reagent failed to show bFGF in the medium (DelliBovi *et al.*, 1987). Culturing of bovine aortic endothelial cells in the presence od NIH-IgBNM6-1 cells or co-culture of the two cell types did not result in accelerated proliferation of the endothelial cells.

The morphology of NIH-N, NIH-BNM7 and NIH-IgBNM6-1 cell lines is shown in Figure 3. The morphology of cells expressing the native bFGF (Figure 3B) does not differ considerably from the control cells (Figure 3A), although some changes in growth properties were observed. The saturation density is about five fold as compared to the control cells, and the growth rate in 2% serum is about doubled. This is in contrast to the high saturation density reached by the chimeric protein elaborating NIH-IgBNM6-1 cell line which displays a very transformed morphology (Figure 3C).

Addition of bFGF neutralizing anti-serum to NIH-IgBNM6-1 cells does not inhibit the transformed morphology, nor does it significantly inhibit its rate of growth (Figure 4; Kurokawa *et al.*, 1989). This observation further suggests that bFGF is not secreted into the medium of the chimeric sp-bFGF producing cells.

The tumorigenicity of the cell lines described here in detail, as well as others like them, was tested in the immunoincompetent nude (Nu/Nu) mice, as well as the immunocompetent syngeneic NFS/NCI mice. The results are shown in Table 1.

Seven differrent bFGF-expressing cell lines gave rise to only one slow-growing tumor among the 30 injected syngeneic mice. Tumor produc-

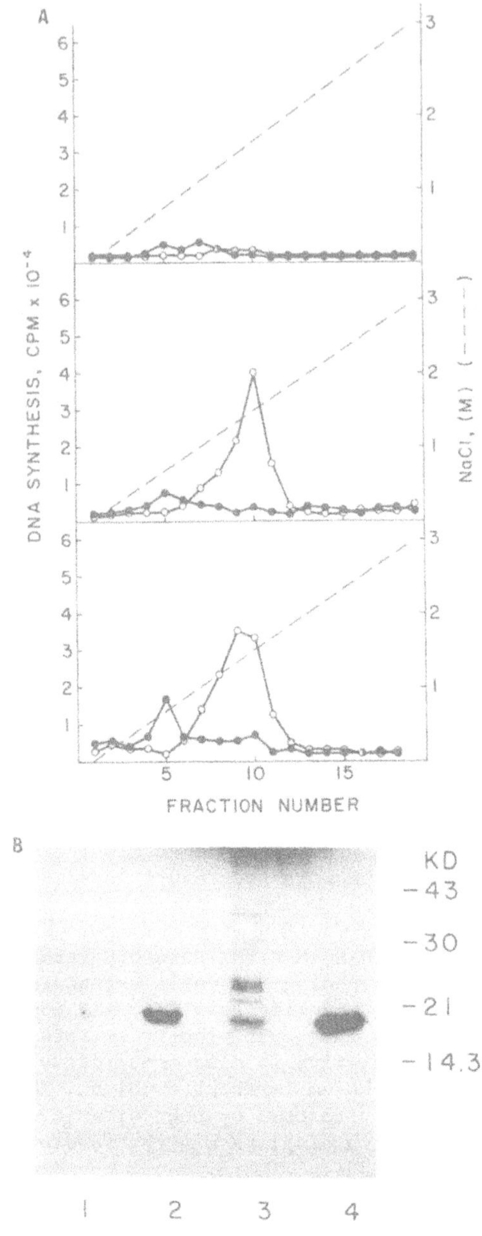

Figure 2. Heparin affinity chromatography and immunoblot analysis of NIH-N, NIH-BNM7 and NIH-IgBNM6-1 cell lines. A. The three panels show Heparin-Sepharose elution profiles. TOP. NIH-N control cells; MIDDLE: NIH-BNM7 cell line expressing the native bFGF; BOTTOM: NIH-IgBNM6-1 cell line expressing the chimeric sp-bFGF protein. Open circles show cell lysates, closed circles show conditioned media. B. Western blot analysis of the 1.5 M NaCl fractions from the above Heparin-Sepharose column. Lane 1: NIH-N control cell lysate; Lane 2: NIH-BNM7 cell lysate; Lane 3: NIH-IgBNM6-1 cell lysate; Lane 4: control recombinant bFGF.

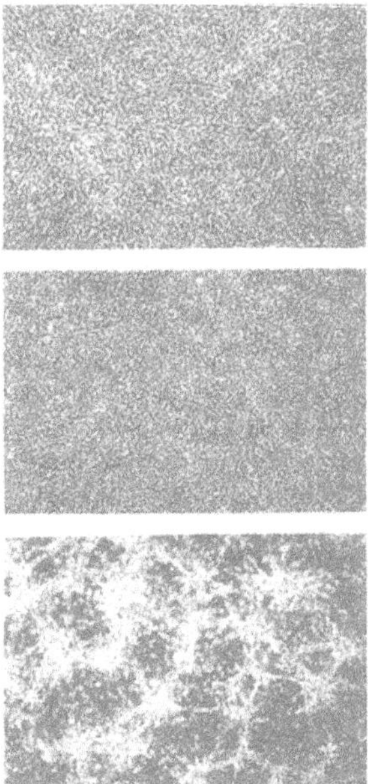

Figure 3. Photomicrographs of TOP: NIH-N control cell line; MIDDLE: NIH-BNM7 cell line expressing the native bovine brain bFGF; BOTTOM: NIH-IgBNM6-1 cell line expressing the chimeric sp-bFGF protein.

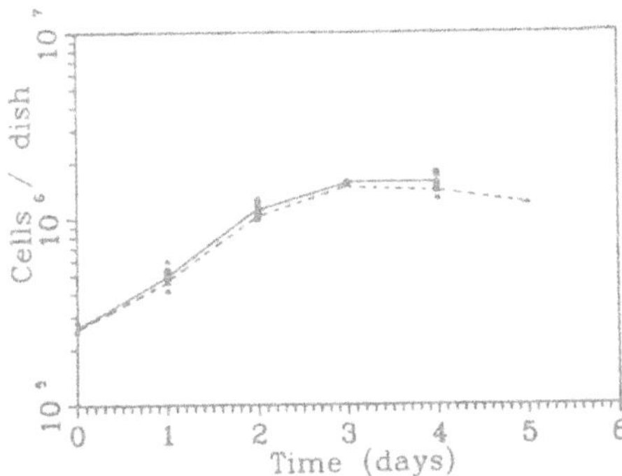

Figure 4. Growth curve of NIH-IgBNM6-1 cell line in the absence (solid line) and in the presence (dashed line) of bFGF neutralizing antibody.

Table 1. Tumorigenicity of NIH-3T3 cells expressing either the native bFGF or the chimeric sp-bFGF.

NIH CELL LINES	U bFGF[1]	Tumors per NCI/NSF mice	Time[2]	Tumors per NUDE/NUDE mice	Time[2]
3T3	0.06	0/12	0/4mo	0/6	0/3mo
BNMpc[3]	0.45	0/6	0/4mo	0/6	0/3mo
BNM6	0.60	0/6	0/3mo	0/6	0/3mo
BNM7	0.80	1/6	5wk/3mo	4/6	3wk/3mo
BNM46	1.40	0/3	0/5mo	2/4	5wk/3mo
BNM48	1.40	0/3	0/5mo	1/4	5wk/3mo
BNM39	3.80	0/3	0/5mo	3/4	5wk/3mo
BNM35	4.60	0/3	0/5mo	3/4	5wk/3mo
spBNM6-1	0.80	38/38	1-2wk	6/6	1-2wk
spBNM4-2	0.70	6/6	1-2wk	4/4	1-2wk
spBNM29-1	0.70	6/6	1-2wk	4/4	1-2wk
EJ 62 Bam2a[4]	0.1	6/6	1-2wk	3/3	1-2wk

[1]Units of bFGF 3T3-mitogenic activity per 10^4 lysed cells.
[2]X/Y where X is time of appearance of tumors about $0.5cm^3$ in size and Y is time during which mice were observed for tumor formation.
[3]A polyclonal cell line expressing the native bFGF (NIH-BNMpc).
[4]An NIH 3T3 cell line transformed with the Ha-ras oncogene.

tion was not dependent on the amount of bFGF produced by the injected cells. However, when these cell lines were injected into immuno-incompetent mice, slow growing tumors arose in 13 out of 34 mice. In fact, the tumorigenic potential in nude mice appeared to be roughly proportional to the level of expression of the native bFGF. It is possible that this low frequency tumor formation in nude muce is due to leakage of bFGF from the dying cells, resulting in paracrine growth stimulation of adjacent cells.

In contrast, NIH-3T3 cells expressing the chimeric sp-bFGF protein gave rise to tumors in 100% nude and 100% syngeneic mice within two weeks after injection. The growth rate of these tumors was comparable to those produced by Ha-ras transformed NIH-3T3 cells. Thus, a relatively low level of of the chimeric sp-bFGF protein is sufficient to induce tumorigenicity, while even five times higher level of the native bFGF does not lead to tumor formation in the syngeneic mice.

In order to determine whether the endogenously produced bFGF was binding to bFGF receptors on the cell surface, NIH-N, NIH-BNM7, NIH-BNM35 (cells expressing about five times as much native bFGF as NIH-BNM7 cells) and NIH-IgBNM6-1 cells were incubated in the presence of ^{125}I bFGF. Closely associated proteins were covalently cross-linked with disiccini-midyl suberate (DSS, Pierce), and equal amounts of cell lysate resolved by SDS--PAGE electrophoresis (Figure 5; Neufeld and Gospodarowica, 1986). Three bFGF binding proteins which, when covalently bound to bFGF migrate at about 180, 170 and 150 Kd are found expressed by the control NIH-N cells. These proteins are also available for binding of the exogenous, radiolabelled bFGF by the native bFGF producing NIH-BNM7 and NIH-BNM35 cell lines. The observed complexes may represent independent receptors of 160, 150 and 130 Kd, or may represent proteolyzed fragments of the

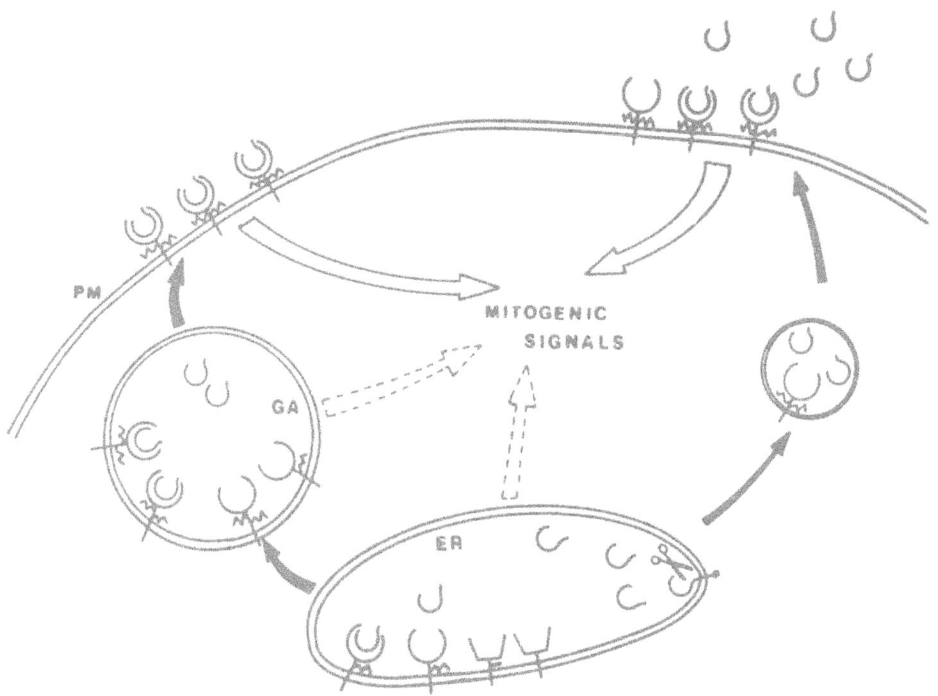

Figure 5. Cell surface receptors affinity labeled with bFGF. Whole cells were incubated in the presence of 10ng/ml of 125-I bFGF for 3 hours on ice. After closely associated proteins were cross-linked with 0.15 mM DSS, the cells were solubilized and equal aliquots were electrophoresed on SDS-PAGE. Lane A: NIH-N control cell line; Lane B: NIH-BNM7 cell line expressing the native bFGF; Lane C: NIH-IgBNM6-1 cell line expressing the chimeric sp-bFGF protein; Lane D: NIH-BNM35 cell line expressing very high levels of the native bFGF.

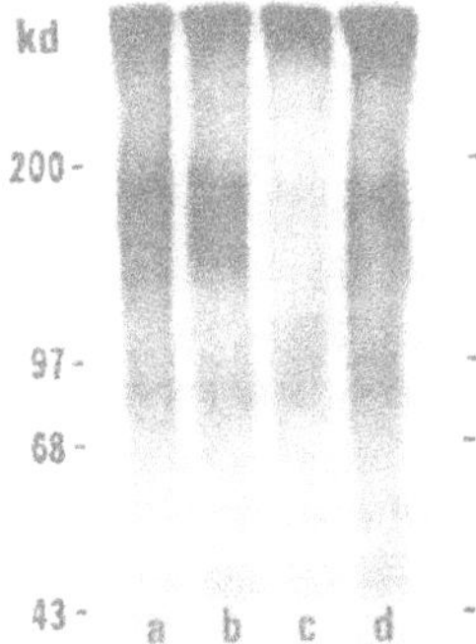

Figure 6. A schematic representation of the different possible modes of autocrine transformation by the chimeric signal peptide-bFGF molecule. RIGHT: Secretion of signal peptide-bFGF precedes binding to the bFGF receptor. The activation of the mitogenic pathway occurs at the plasma membrane. LEFT: signal peptide-bFGF chimeric molecule already binds to the receptor in the endoplasmic reticulum or the Golgi. The mitogenic pathway is activated after the complex reaches the cell surface or perhaps even directy from the endoplasmid reticulum or the Golgi apparatus.

true bFGF receptor protein(s). Few of these bFGF binding proteins appear available on the surface of sp-FGF producing NIH-IgBNM6-1 cells. The receptors on these cells may be fully occupied with endogenously produced bFGF, or may be down-regulated. The decrease in free receptors appears to be primarily due to down-regulation, since washing the cell surface with 2M NaCl in 20 mM.

We conclude that expression of bFGF by NIH-3T3 fibroblasts yields a differential phenotype depending on whether the protein has a signal peptide. The native bFGF molecule does not endow the cells with transformed phenotype, while addition of a signal peptide to this growth factor results in autocrine transformation.

A schematic representation of the model for the autorine transformation by bFGF is shown in Figure 6. The native bFGF molecule lacking a signal peptide may remain trapped in the cytoplasm of the producing cell, physically separated from the bFGF receptors on the cell surface and in the endoplasmic reticulum and Golgi apparatus. Thus, such a gowth factor has no effect on the cell's growth. In spite of the lack of bFGF in the conditioned medium, the presence of cleaved 18 Kd protein in sp-bFGF producing cells suggests that the chimeric growth factor has entered the endoplasmic reticulum. Thus bFGF is localized to the same biosynthetic pathway that is used by the receptors for the known growth factors as they travel to the cell surface. Although it is possible that bFGF is secreted into the medium and is immediately adsorbed to the cell surface receptors, it is more likely that bFGF binds to its receptor along the route to externalization and reaches the cell surface as a complex. At that site, if not earlier, the complex may activate the mitogenic pathway. This continuous activation may indeed cause cell surface receptor down-regulation, but since the growth factor approaches the receptor intracellularly, the autocrine growth stimulation continues, and results in the transformed phenotype with tumorigenic potential.

REFERENCES

Abraham, J. A., Mergia, A., Whang, J.L., Tumolo, A., Friedman, J., Hjerrild, K.A., Gospodarowicz, D. and Fiddes, J.C. 1986. Nucleotide sequence of a bovine clone encoding the angiogenic protein, basic fibroblast growth factor. Science, 233:545

DelliBovi, P., Curatola, A.M., Kern, F.G., Greco, A., Ittmann, M. and Basilico, C. 1987. An oncogene isolated by transfection of Kaposi Sarcoma DNA encodes a growth factor that is a member of the FGF family. Cell, 50:729

Doolittle, R.F., Hunkapiller, M.W., Hood, L.E., Devare, S.G., Robbins, K.C., Aaronson, S.A. and Antonaides, H.N. 1983. Simian sarcoma virus oncogene, v-sis, is derived from the gene (or genes) encoding a platelet derived growth factor. Science, 221:275

Finzi, E., Fleming, T., Segatto, D., Pennington, C.Y., Bringman, T.S., Derynck, R. and Aaronson, S.A. 1987. A human transforming growth factor type alpha coding sequence is not a direct-acting oncogene when overexpressed in NIH3T3 cells. Proc. Natl. Acad. Sci.(USA), 184:3733

Folkman, J. and Klagsbrun, M. 1987. Angiogenic factors. Science, 235:442

Hannik, M. and Donoghue, D. J. 1988. Autocrine stimulation by the v-sis gene product requires a ligand-receptor interaction at the cell surface. J. Cell Biol., 107:287.

Keating, M.T. and Williams L.T. 1987. Autocrine stimulation of intracellular PDGF receptors in v-sis transformed cells. Science, 239:914.

Klagsbrun, M., Sasse, J., Sullivan, R. and Smith, J.A. 1986. Human tumor cells synthesize an endothelial cell growth factor that is structurally related to basic fibroblast growth factor. Proc. Nat. Acad. Sci.(USA), 83:2448

Kurokawa, M., Doctrow, S. and Klagsbrun, M. 1989. Neutralizing antibodies inhibit binding of bFGF to its receptor but not to heparin. J. Biol. Chem., In press.

Lang, R.A., Metcalf, D., Gough, N.M., Dunn, A.R. and Gonda, T.J. 1985. Expression of a hemopoietic growth factor cDNA in a factor-dependent cell line results in an autonomous growth and tumorigenicity. Cell, 43:531

Leal, F., Williams, L.T., Robbins, K.C. and Aaronson, S.A. 1985. Evidence that the v-sis gene product transforms by interaction with the receptor for platelet-derived growth factor. Science, 230:327

Loh, D.Y., Bothwell, A.L.M., White-Scharf, M.E., Imanishi-Kari, T. and Baltimore, D. 1983. Molecular basis of a mouse strain-specific anti-hapten response. Cell, 33:85

Neufeld, G. and Gospodarowicz, D. 1986. Basic and acidic growth factors interact with the same cell surface receptor. J. Biol. Chem., 261:5631

Robbins, K.C., Leal, F., Pierce, J.H. and Aaronson, S.A. 1985. The v-sis/PDGF-2 transforming gene product localizes to cell membranes but is not a secretory protein. EMBO J., 4:1783

Rogelj, S., Weinberg, R.A., Fanning, P. and Klagsbrun, M. 1988. Basic fibroblast growth factor fused to a signal peptide transforms cells. Nature (London), 331:173

Rosenthal,A., Lindquist, P.B., Bringman, T.S., Goeddel, D.V. and Derynck, R. 1986. Expression in rat fibroblasts of a human transforming growth factor alpha cDNA results in transformation. Cell, 46:301

Schweigerer, L., Neufeld, G., Friedman, J., Abraham, J.A., Fiddes, J.C. and Gospodarowicz, D. 1987. Capillary endothelial cells expresss basic fibroblast growth factor, a mitogen that promotes their own growth. Nature (London), 325:257

Southern, P. J. and Berg, P. 1982. Transformation of mammalian cells to antibiotic resistance with a bacterial gene under control of the SV40 early region promoter. J. Mol. Appl.Genet., 1:327

Sporn, M. B. and Todaro, G. J. 1980. Autocrine secretion and malignant transformation of cells. N. Engl. J. Med. 303:878

Stern, D. F., Hare, D.L., Cecchin, M.A. and Weinberg, R.A. 1987. Construction of a novel oncogene based on Synthetic Sequences Encoding Epidermal Growth Factor. Science, 235:321

Vlodavsky,I., Folkman, J., Sullivan, R., Fridman, R., Ishai-Michaeli, R.,

Sasse, J. and Klagsbrun, M. 1987. Endothelial cell-derived basic fibriblast growth factor: synthesis and deposition into subendothelial extracellular matrix. Proc. Natl. Acad. Sci.(USA), 84:2292

Watanabe, S., Lazar, E. and Sporn, M.B. 1987. Transformation of normal rat kidney (NRK) cells by an infectious retrovirus carrying a synthetic rat type alpha transforming growth factor. Proc. Natl. Acad. Sci. (USA), 84:1258

Waterfield, M.D., Scarce, G.T., Whittle, N., Stroobert, P., Johnsson, A., Wasteson, A., Westermark, B., Heldin, C-H., Huang, J.S. and Deuel, T.F. 1983. Platelet-derived growth factor is structurally related to the putative transforming protein of Simian sarcoma virus. Nature (London), 304:35

Wheeler, E.F., Rettenmier, C.W., Look, A.T. and Sherr, C.J. 1986. The v-fms oncogene induces factor independence and tumorigenicity in a CSF-1 dependent macrophage cell line. Nature (London), 324:377

PROSTATE GROWTH FACTOR: CHARACTERIZATION AND ITS ROLE IN NORMAL PROSTATE AND IN BENIGN PROSTATIC HYPERPLASIA

Michael T. Story

Departments of Urology and Biochemistry
Medical College of Wisconsin
9200 West Wisconsin Avenue
Milwaukee, WI 53226

Benign prostatic hyperplasia (BPH) is the most common benign neoplastic growth that occurs in men. It has been estimated that 75% of males over 50 years of age have symptoms of urinary-outlet obstruction arising from BPH (Walsh, 1984). Birkhoff (1983) has estimated that the chance of a 50 year old requiring a prostatectomy during his lifetime is 20-25%. Over 350,000 surgical procedures are performed each year to relieve symptoms resulting from BPH, resulting in an overall medical cost that exceeds one billion dollars annually in the United States (Peters and Walsh, 1987). These statistics emphasize the need for new approaches to the study of BPH because, despite more than 100 years of study, the etiology of BPH has not been resolved.

The age-related changes in the size of the prostate and the incidence of BPH has been compiled by Walsh (1984). The data demonstrate that there is a slow increase in the size of the prostate from birth until puberty. At puberty, a rapid increase in growth occurs that continues until about age 30. The size of the prostate then remains constant until about age 45. Thereafter, there is an increase in prostate weight that is accompanied by an increased incidence of BPH. Histological evidence of BPH increases from an average of 23% in the forties to 88% by age 90. In addition, the amount of tissue removed at prostatectomy increases with age. These data support the progressive nature of the disease in the aging male.

The importance of the testes in the development and maintenance of prostatic function in animals has been recognized since the time of Hunter dating from the 18th century (Hunter, 1841). It is known that the testes must be present for development and maintenance of the secretory function of the human prostate and that BPH does not develop in men castrated before puberty (Moore, 1944). The evidence also indicates that androgens are necessary to maintain BPH, but a causal relationship between steroid hormone and the development of the human disease has not been established (see review by Walsh, 1984).

There is general agreement that BPH begins in the inner region of the prostate adjacent to the urethra between the verumontanum and the bladder neck. The close proximity of the hyperplastic growth to the urethra accounts for the high incidence of urinary obstruction in BPH.

Reischaurer (1925) described the first lesion of BPH as a fibromyomatous nodule. He stated that the stromal nodules induced a very early and rapid penetration of glandular elements. Subsequent morphologic studies by Deming and Neumann (1939), LeDuc (1939), and Moore (1943) confirmed this thesis. A more recent report by McNeal (1984) described the cells of the periurethral, predominant stromal, nodule as resembling embryonic mesenchyme, with an abundance of pale ground substance and only a few fine collagen fibers. Most transition zone nodules, on the other hand, appeared glandular with little stroma. With advancing age, glandular and stromal nodules increase in number, but large nodules are uncommon until well into the seventh decade. In advanced BPH there is a marked coordinated enlargement of most of the nodules within the prostate. The true prostatic tissue is displaced caudad and posteriorly as the lobes of benign hyperplasia grow and distort the original anatomy of the gland. The lateral and median lobes seen in patients with BPH are masses of hyperplastic tissue that have displaced the true prostatic tissue to the periphery (Lawson, 1986).

McNeal (1984) suggested that the genesis of the BPH nodule results from a disturbance arising at the tissue level within the susceptible region of the prostate that results in the reversion of a clone of stroma cells to the embryonic state. The foci of embryonic stroma then induces the growth of adjacent epithelium giving rise to the glandular nodule. The duct formation and epithelial hyperplasia appeared greater on the wall facing the embryonic stroma suggesting that the inductive influence of the stroma may be mediated by diffusible inducers.

McNeal's hypothesis was based entirely on his extensive morphologic observations of the prostate. The test of the hypothesis will require the application of protein purification techniques to the isolation of hypothetical stroma inducers and the techniques of modern molecular biology to probe for genes responsible for embryonic reawakening in the adult prostate.

ISOLATION OF PROSTATE GROWTH FACTOR AND IDENTIFICATION AS BASIC FIBROBLAST GROWTH FACTOR

In 1979 our laboratory reported that extracts of human prostate were mitogenic for cultured cells (Jacobs *et al.*, 1979). Cells responding to prostatic extracts included osteoblasts derived from fetal rat calvaria, fetal rat skin fibroblasts, human foreskin fibroblasts and fibroblasts derived from BPH tissue (Table 1). Extracts from normal post-pubertal prostate, BPH prostate and well differentiated prostate cancer were found to stimulate ^{3}H-thymidine incorporation by the cultured cells; whereas, extracted of human muscle and kidney contained little if any mitogenic activity. The growth-stimulating factor was both heat and trypsin sensitive and was non-dialyzable indicating that the factor was likely a protein (Jacobs and Lawson, 1980). At that time, we postulated that prostate contained a protein growth factor that may function in the development of BPH and in the osteoblastic response to prostate cancer. The next several years were spent in isolating and characterizing the growth factor from BPH tissue.

To accomplish this objective, it was necessary to have a reliable assay for the growth factor. The assay we developed (Story *et al.*, 1983) is shown in Figure 1. Human foreskin fibroblasts, passage 2 through 8, were plated in 24-well culture chambers in medium containing 10% newborn bovine serum-(NBS). Cells were grown to confluency and the cells were washed and incubated 1 day in medium with 0.5% serum. Triplicate wells

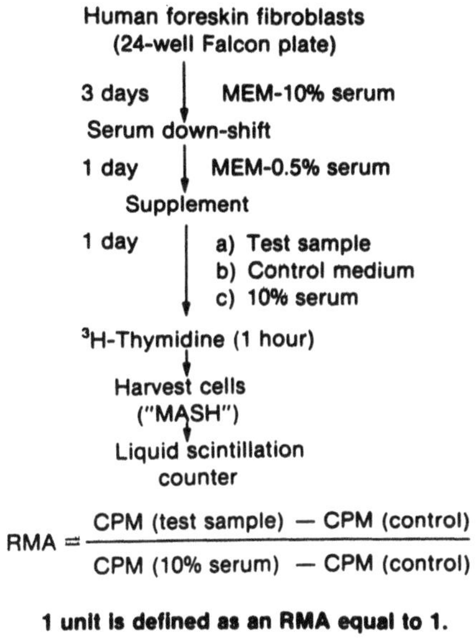

Human foreskin fibroblasts
(24-well Falcon plate)

3 days ↓ MEM-10% serum

Serum down-shift

1 day ↓ MEM-0.5% serum

Supplement

1 day ↓ a) Test sample
b) Control medium
c) 10% serum

^{3}H-Thymidine (1 hour)

↓

Harvest cells
("MASH")

↓

Liquid scintillation
counter

$$\text{RMA} = \frac{\text{CPM (test sample)} - \text{CPM (control)}}{\text{CPM (10\% serum)} - \text{CPM (control)}}$$

1 unit is defined as an RMA equal to 1.

Figure 1.

of cells received 10% NBS (positive control), samples to be tested for growth factor activity, and buffer in which the test samples were prepared (negative control). One day later, a time of peak thymidine incorporation into DNA (Jacobs and Lawson, 1980, Lawson *et al.*, 1981) cells were pulsed 1 hr with 3H-thymidine. The cells were washed with non-radioactive buffer and collected on a glass fiber filter, and the amount of radiolabel incorporated was determined.

Human prostate tissue was from surgical specimens obtained at transurethral (TURP) and open prostatectomy for BPH. As seen in Table 2

Table 1. Cells responding to prostatic extract

Extract	Fetal Rat Osteoblasts	Fetal Rat Skin Fibroblasts	Human Foreskin Fibroblasts	Human BPH-derived Fibroblasts
BPH	550 ± 28	586 ± 25	254 ± 6	419 ± 22
Normal Post Pubertal Prostate	385 ± 15	328 ± 19	313 ± 22	ND
Adenocarcinoma Prostate (Lymph Node)	ND	459 ± 19	ND	ND
Muscle	186 ± 28	130 ± 9	109 ± 6	ND
Kidney	122 ± 10	131 ± 19	ND	ND

Values are the percent of control ± S.D.; ND = not determined.

Table 2. Growth factor activity of BPH homogenates prepared at pH 7.6 in low and high ionic strength buffer and at pH 4.5[a]

	extraction conditions		Activity	
source of BPH tissue	ionic strength	pH	units/mg of protein	units/g wet wt
open prostatectomy 420 (300-890)	low[b]	7.6	20 (10-47)	
open prostatectomy (2758-3086)	high[b]	7.6	143 (112-165)	2949
TURP (1198-2568)	high[c]	7.6	151 (101-234)	1761
TURP (48-375)	-[d]	4.5	161 (87-344)	217

[a]Mean and range ($N \leq 4$). [b]50 mM Tris/50 mM NaCl. [c]50 mM Tris/1.55 M NaCl with EDTA (10 mM), phenylmethanesulfonyl fluoride (1 mM), L-1-(tosylamido)-2-phenylethyl chloromethyl ketone (0.03 mM), ethylmaleimide (0.05 mM), and soybean trypsin inhibitor (10 mg/L). [d]As described by Gospodarowicz *et al.*, (1985a). From Story *et al.*, (1987b) with permission.

Table 3. Isolation scheme for PrGF

STEPS	PROCEDURES
1.	Homogenization in high ionic strength buffer with protease inhibitors.
2.	Dichloromethane extraction (aqueous phase).
3.	Ammonium sulfate precipitation, 30% saturation (supernatant).
4.	Phenyl-Sepharose (gradient elution with ethylene glycol).
5.	Ultrafiltration.
6.	G-75, low ionic strength buffer.
7.	DEAE (elution with high ionic strength buffer).
8.	Ultrafiltration.
9.	G-75, high ionic strength buffer.
10.	Con A-Sepharose (unbound fraction).
11.	Ultrafiltration.
12.	P-150, high ionic strength buffer.
13.	Ultrafiltration.

Table 4. Application of the purification scheme to three BPH preparations

Study No.	Gr of BPH Tissue	Fold Increase in Activity	Final Growth Factor Activity	Percent Recovery
1	103	1400	236,250 (4ng = 1 unit)	1
2	101	1000	55,143 (18ng = 1 unit)	6
3	100	500	55,000 (18ng = 1 unit)	14

Initial yield = 32.5%; average repetitive yield - 89.9%
[a]Phenyltiohydantoin amino acid derivative
[b]Tentative but probable amino acid identified
From Story *et al.*, (1987b) with permission.

(Story *et al.*, 1987b), Polytron homogenates of tissue obtained by open prostatectomy contained on the average 420 units of growth factor activity per g (wet weight) when homogenized in low ionic strength buffer at pH 7.6. However, when homogenized in high ionic strength buffer (1.5 M NaCl) with protease inhibitors, an average of 2949 activity units/g was present. Homogenates of TURP specimens had somewhat lower activity, 1761 units/g, when homogenized by similar procedures, but when homogenized at acid pH (pH 4.5), only 217 units/g. Thus, prostate growth factor (PrGF) appeared acid labile, and the amount of activity recovered was greatly enhanced by homogenizing the tissue in high ionic strength buffer. Prostate growth factor activity was inactivated by urea and guanidine-hydrochloride at neutral pH (Story *et al.*, 1984b) but was stable to thio-reducing agents. The growth factor activity of homogenates was precipitated between 33 and 67% ammonium sulfate, remained in the aqueous phase after either or dichloromethane extraction and did not bind to concanavalin A. These properties were consistent with the protein nature of the growth factor.

The growth factor activity of prostate homogenates eluted from gel filtration columns, run in low ionic strength buffer, with an apparent molecular weight of (M_r) > 67,000. However, when fractionated in high ionic strength buffer, activity eluted with an apparent M_r of 17,000 (Story *et al.*, 1984a). The apparent isoelectric point (PI) of the partially purified growth factor, which was also determined at low ionic strength, indicated an acidic PI.

To summarize, PrGF appeared to be an acidic single-chain polyeptide of M_r 17,000 that was heat and acid labile. These properties of the growth factor, and the shift in apparent M_r as a function of ionic strength, were utilized in a 13 step isolation scheme for PrGF (Table 3). The scheme resulted in a 500- to 1400-fold purification of the growth factor but recovery was only 1 to 14 percent (Table 4). The growth factor preparations were contaminated with a number of proteins that were evident when electrophoresed on polyacrylamide gels.

Shing and co-workers (1984) reported the isolation of chondrosarcoma-derived growth factor from a rat transplantable tumor by a two step procedure of cation-exchange and heparin-affinity chromatography.

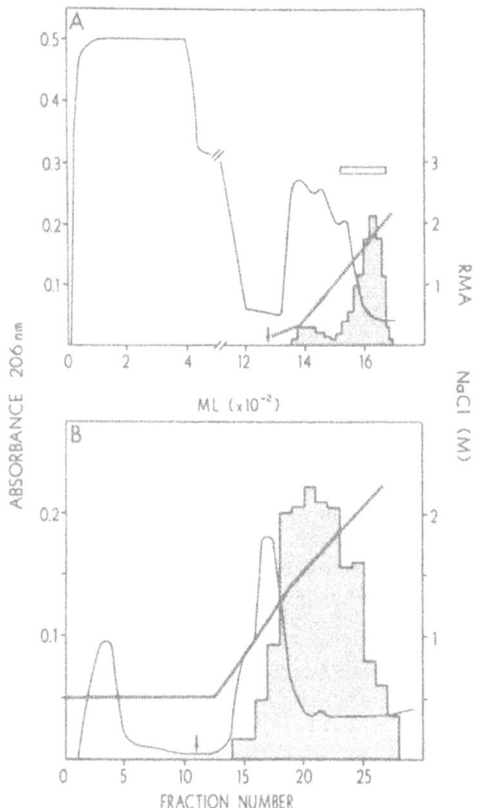

Figure 2. Heparin-Sepharose Chromatography. (A) BPH tissue (361 g) was homogenized in 50 mM Tris/1.55 M NaCl containing protease inhibitors, pH 7.6, precipitated with ammonium sulfate (25-75% saturation, pH 7.6), and dialyzed against 50 mM Tris/0.15 M NaCl, pH 7.6 (TBS). Half of the sample was applied to a column of heparin-Sepharose (5 cm x 2 cm, 40 ml bed volume) equilibrated with TBS. The column was washed with 870 ml of TBS, and a gradient (480 ml) of TBS with 3 M NaCl was started (arrow). Fractions were collected, and the absorbance (-), conductivity, and growth factor activity (shaded area) were measured. The activity eluting between 1.2 and 2.1 M NaCl (bar) from two 40 ml columns were concentrated, dialyzed against TBS with 0.5 M NaCl, and applied to a 1.6 cm x 7.5 cm, 15 ml bed volume, column of heparin-Sepharose (B) equilibrated with TBS/0.5 M NaCl. The column was washed with equilibration buffer until the absorbance reached base line. A gradient (280 ml) of TBS/0.5 M NaCl to TBS with 3 M NaCl was started (arrow). Fractions (12 ml) were collected, and the absorbance, conductivity, and growth factor activity were measured. The legends are as in (A). Fractions 18-25 were dialyzed against 50 mM Tris/50 mM NaCl, pH 7.6, concentrated, and analyzed by SDS-PAGE (Figure 3, lane 3). From Story *et al.* (1987b) with permission).

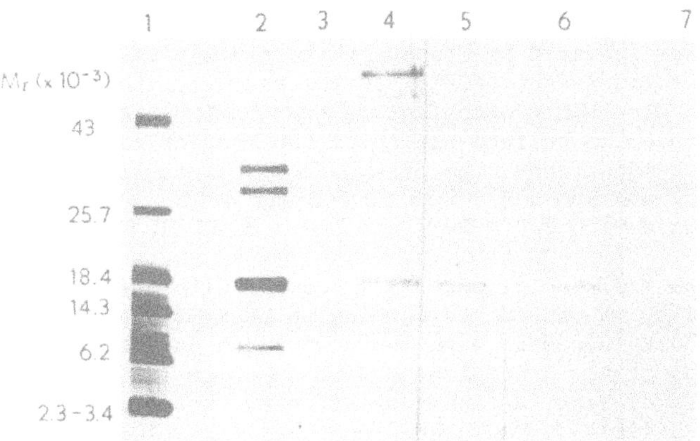

Figure 3. SDS PAGE of PrGF and Immunodetection by Anti-bFGF Antisera. A preparation of PrGF partially purified by two cycles of heparin-Sepharose chromatography were electrophoresed by SDS-PAGE, Laemmli system with a 10-22.5% acrylamide gradient. The proteins were transferred electrophoretically in replicate to nitrocellulose paper. The SDS-PAGE gel was silver stained for protein (lanes 1-4), and the nitrocellulose-transferred protein was incubated with antisera against synthetic peptides with sequence homologies to bovine bFGF. Bound antibody was visualized by incubation with biotinylated peroxidase complex (lanes 5-7). Lane 1, M_r markers; lane 2, heparin-Sepharose partially purified pRGF; lane 3, sample buffer; lane 4, cation-exchange purified PrGF. Lanes 5-7, heparin-Sepharose partially purified PrGF incubated with antisera to bFGF peptides: Lane 5, bFGF(33-43); lane 6, bFGF(1-12); lane 7, bFGF(136-145). From Story *et al.*, (1987b) with permission.

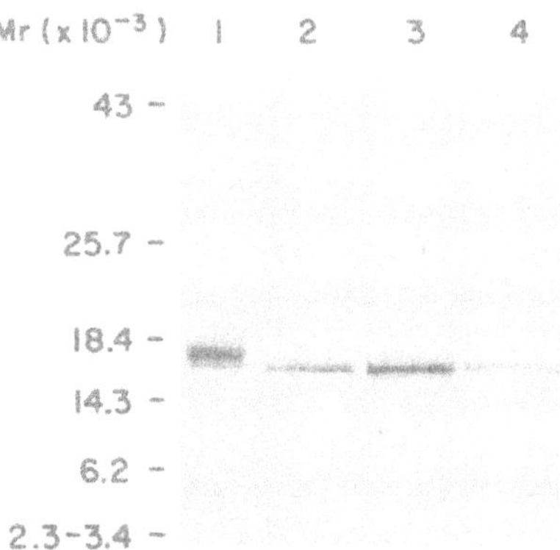

Figure 4. Detection of Growth Factor by Antisera to Synthetic Peptides with Sequence Homologies to bFGF. Approximately 4,000 units of growth factor was applied per lane, electrophoresed, transferred to nitrocellulose paper, and incubated with antisera to synthetic peptides with sequence homologies to bFGF. Bound antibody was visualized by incubation with peroxidase conjugated goat-anti-rabbit IgG. The position of M_r markers, visualized by silver staining of the original gel, is indicated. Lane 1, growth factor purified from alkaline extracts: Lanes 2-4, growth factor isolated from acidic extracts. Lanes 1-2, antiserum to bFGF (1-12); lane 3, antiserum to bFGF (33-43); lane 4, antiserum to bFGF (136-145). From Story *et al.*, (1987a) with permission.

The growth factor, for capillary endothelial cells and 3T3 cells, had a PI of 9.5-10 and a M_r of 18,000. The growth factor had properties similar to one isolated by Gospodarowitz (1974) from bovine brain, termed fibroblast growth factor (FGF). In the same year Lobb and Fett (1984) purified two distinct growth factors from bovine brain by homogenization at pH 4.5, ammonium sulfate precipitation, cation-exchange and heparin-Sepharose chromatography. One growth factor eluted from heparin-Sepharose with 1 M NaCl and had a M_r of 16 KD and was acidic. The second peak of activity eluted with 2.0 M NaCl and had a M_r of 18 KD.

Because PrGF had properties in common with the heparin-binding growth factors, we tested its binding to heparin-Sepharose. The BPH tissue was homogenized in high ionic strength buffer (pH 7.6), containing protease inhibitors, precipitated with ammonium sulfate and fractionated by two cycles of heparin-Sepharose affinity chromatography (Figure 2, Story *et al.*, 1987b). The predominant activity eluted from the second heparin column between 1.4 and 2 M NaCl. The growth factor activity was enhanced 8,488 fold and recovery was 34 percent. The preparation when electrophoresed on SDS-PAGE gels contained a predominant protein of M_r 17,600 but also contained other less prominent proteins (Figure 3, Lane 2). The 17,600 M_r component was further purified by cation-exchange chromatography. Unlike partially purified PrFG which appeared to have an acidic PI, the purified growth factor had a PI of 10.2. In collaboration with Dr. Joachim Sasse, (Harvard Medical School), who provided antisera to synthetic basic FGF (bFGF) synthetic peptides, the immunologic identity of PrGF and bFGF was confirmed (Figure 3, Lanes 4-7).

Bovine bFGF had been sequenced by Esch *et al.* (1985) and was reported to be a 246 amino acid polypeptide. However, the 17,600 M_r growth factor that we isolated appeared larger than the 16,415 M_r polypeptide determined by sequence analysis. When the alkaline extracted growth factor was electrophoresed side-by-side with the growth factor recovered from acidic extracts of BPH tissue, the difference in M_r of the two growth factors was evident (Figure 4, Lanes 1 and 2). The M_r of the alkaline extracted growth factor was estimated to be 17,400 and the acidic extracted growth factor was 16,600. Both growth factors reacted with antisera to synthetic peptides with sequences corresponding to the amino-terminal (1-12), internal (33-43) and carboxyl-terminal (136-145) amino acid residues of bovine bFGF (Figure 4). This suggested that the smaller acidic extracted growth factor had both amino-and carboxyl-terminal ends. Therefore, it was likely that the larger alkaline extracted growth factor had an additional sequence not present in (1-146) bFGF. In collaboration with Dr. F. Esch, the alkaline extracted PrGF was found to contain an amino-terminal 8 amino acid extension not found in (1-146)bFGF (Table 5, Story *et al.*, 1987a). The extended form of the molecule had the sequence, Ala-Ala-Gly-Ser-Ile-Thr-Thr-Leu. This sequence agreed with the sequence of 8 of the 9 extra residues at the amino-terminus of bFGF predicted from the nucleotide sequence of a bovine brain clone encoding bFGF that was reported by Abraham *et al.* (1986). The amino-terminal methionine residue was not present. The evidence indicated that the form of bFGF that exists in BPH tissue is (-8-146) and that variations in the form of the growth factor isolated are related to the homogenization procedure. Homogenization of BPH tissue in high ionic strength alkaline buffer containing protease inhibitors leads to the isolation of (-8-146)bFGF. When BPH tissue is homogenized at pH 4.5, without protease inhibitors, the yield of growth factor is greatly reduced (Story *et al.*, 1987b) and a form that lacks the amino-terminal extension, likely (1-145), is recovered (Figure 4). Both the (-8-146) and (1-146) forms of bFGF were also isolated from bovine pituitary at pH 4.5 in the presence of protease inhibitors (Ueno *et al.*, 1987). Thus,

Table 5. Amino terminal sequence analysis of prostatic growth factor (67 pmol)

Residue	>PnNCS-AA[a]	(pmol)
1	Ala	21.8
2	Ala	19.9
3	Gly	9.0
4	Ser	6.1
5	Ile	8.4
6	Thr	3.5
7	Thr	3.0
8	Leu	5.4
9	Pro	2.9
10	Ala	5.3
11	Leu	3.8
12	Pro	2.9
13	Glu	3.6
14	Asp	1.1
15	Gly	2.3
16	(Gly)[b]	-
17	Ser	0.8
18	(Gly)[b]	-
19	(Ala)[b]	-
20	Phe	1.3
21	Pro	1.5

Initial yield = 32.5%; average repetitive yield - 89.9%
[a]Phenyltiohydantoin amino acid derivative
[b]Tentative but probable amino acid identified
From Story *et al.*, (1987b) with permission.

isolation of the amino-terminal extended form of the growth factor is not strictly related to pH during homogenization. It's likely that the (1-146) form was due to proteolysis that was, in part, prevented protease inhibitors in the homogenization buffer. A (16-146)bFGF form isolated from tissues at pH 4.5, without protease inhibitors, (Gospodarowicz *et al.*, 1985; Gospodarowicz *et al.*, 1986; Baird *et al.*, 1985) seems to be a result of tissue specific enzymes, activated during extraction, which further degrade the growth factor.

The bioassay used to monitor the purification of the growth factor from prostate is not specific for bFGF. In order to study the biologic properties of bFGF *in vivo*, a specific and sensitive assay for bFGF is needed. Ideally, the assay should have the sensitivity to quantitate bFGF in biopsy specimens containing 10 mg of tissue. From the values in Table 2, 10 mg of BPH tissue would be expected to contain about 30 units of growth factor activity. A unit of activity in our assay is equivalent to about 0.7 ng of pure bFGF. It follows that 10 mg of BPH tissue contains about 20 ng of bFGF. Secondly, specificity should permit quantitation of the growth factor in crude BPH tissue homogenates. Thirdly, the assay should detect truncated forms of the growth factor that may be generated during extraction.

Polyclonal antisera have been raised against synthetic bFGF peptides and to the native molecule. These antisera have been valuable in identifying the growth factor in several tissues (Story *et al.*, 1987B;

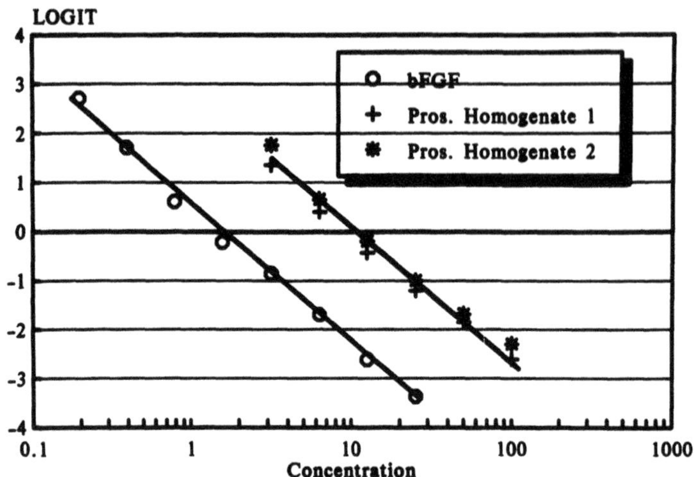

Figure 5. Competition of bFGF and Prostate Homogenates 1 and 2 for Binding of ^{125}I-bFGF to Antiserum generated against (1-24)bFGF-BSA.

Story *et al.*, 1988B; Klagsbrun *et al.*, 1986). Antisera against synthetic bFGF peptides react with reduced bFGF on immunoblots but frequently do not react with the native molecule in solution. If they do react, they may lack the desired specificity. These include, antisera to synthetic bFGF peptides (33-43), (30-50), and (136-145) evaluated by us and antisera to sequence (73-87) and (69-87) studied by Halaban *et al.* (1987). The inability of these antisera to fully react with the native protein suggests that these epitopes are not completely accessible to antibody.

Polyclonal antisera to the amino-terminal synthetic bFGF peptides (1-10), (1-12) and (1-15) have been used in radioimmunoassays (Baird *et al.*, 1985a; 1985b). Baird and co-workers (1985a) reported that bFGF competed for binding with antiserum against (tyr10)bFGF (1-10)-conjugated to bovine serum albumin on an equimolar basis with (1-10)bFGF. However, the antiserum lacked specificity and required partial purification of the growth factor before assay. Furthermore, the peptide was rapidly biodegraded in serum samples (Gauthier *et al.*, 1987), which gave misleading results. Our experience with antiserum against (1-12) bFGF indicated that it could not be used to assay the growth factor in crude tissue homogenates. Dose-response curves of (1-12)bFGF for antibody binding showed competition on an equimolar basis, but the antiserum did not recognize, on an equimolar basis, bFGF in tissue homogenates. This suggested that antibody species were present in the polyclonal antiserum that reacted with similar sequences in tissue proteins. This is not surprising, since the number of epitopes available on small synthetic peptides is limited and the same sequence might be expected to occur in other proteins.

Rabbit antiserum to (1-24)bFGF-BSA was kindly provided by Dr. Andrew Baird (The Salk Institute for Biological Studies). The antiserum was used in a competitive binding assay with recombinant bFGF (Amgen) as standard and trace. The antiserum has the specificity required to quantitate bFGF in crude prostate tissue preparations. This is apparent from the parallel dose-response curves of the competition of bFGF and tissue homogenates for ^{125}I-bFGF binding to antiserum (Figure 5). The assay sensitivity is about 250 pg which permits quantitation of bFGF from

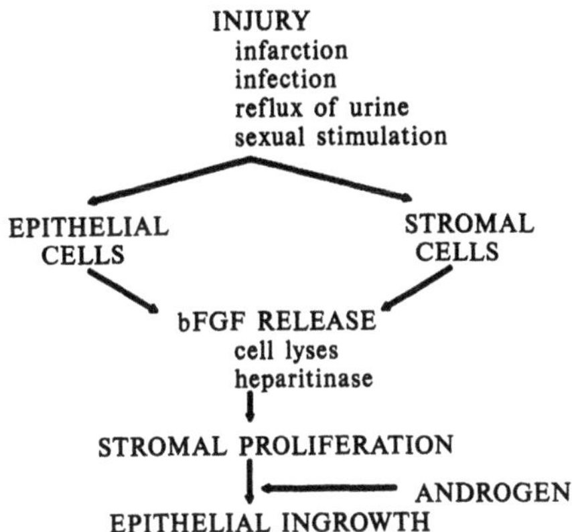

Figure 6. A working model for the role of bFGF in BPH

10 mg of BPH tissue. The sequence of bFGF recognized by Dr. Baird's antiserum is amino acid residues 11-15 (Baird and Ling, 1987). A limitation of the antiserum is the inability of the antibody to recognize amino-terminal truncated forms of bFGF. Multiple forms of bFGF arise from amino-terminal cleavages by acid proteinases during tissue preparation (Baird *et al.*, 1987; Klagsbrun *et al.*, 1987). The form (16-146)bFGF has been isolated from a number of tissues (Gospodarowicz *et al.*, 1985; Ueno *et al.*, 1987; Baird *et al.*, 1985b). This form of the growth factor has full biologic activity (Gospodarowicz, 1985) but since it is lacking amino acid residues 11-15, it is not recognized by Dr. Baird's antiserum. Thus, the level of bFGF in tissue may be underestimated when this antiserum is used for quantitation, particularly when tissue is extracted at acid pH. Monoclonal antibodies to bFGF, recently described by Massoglia and co-workers (1987) and others developed in our laboratory that are currently being characterized, are expected to overcome the limitation of the currently available reagents.

A WORKING MODEL FOR THE ROLE OF BASIC FGF IN BPH

A working model for the role of bFGF in BPH is outlined in Figure 6. We propose that injury damages prostatic epithelial or stromal cells resulting in the release of bFGF. Once released from cells or from extracellular reservoirs, bFGF binds to stroma cell surface receptors resulting in stromal proliferation. The primary stromal nodule induces epithelial ingrowth, that is maintained by androgen, producing the mixed nodules characteristic of BPH (Lawson, 1986; Lawson, 1988).

Is FGF important in the development of BPH?

The role of bFGF *in vivo*, like that of the other growth factors that have been described, is poorly understood. The high degree of structural conservation of bFGF in all of the species in which it has been studied, suggests that it has an important function in regulating growth and development. Recent work identifying bFGF as the major angiogenesic factor in tissue and the finding that it is a potent stimulator of fibrostromal growth has focused attention on this growth factor as a major effector in wound healing and normal tissue repair. These properties of the growth factor are discussed below in greater detail.

Little is known about bFGF expression in neoplastic transformation. Tumor cells are known to produce the growth factor (Shing *et al.*, 1984; Moscatelli *et al.*, 1986). Evidence that a viral oncogene may code for a growth factor or share structural homology with growth factor receptors has emerged from studies of platelet-derived growth factor and epidermal growth factor (Doolittle *et al.*, 1983; Downward *et al.*, 1984). These growth factors have similar mechanisms of action as bFGF. There is little doubt that FGFs are also the product of oncogenes. In fact, the int-2 and hst/ks transforming genes have a high degree of structural homology with FGF (Dickson and Peters, 1987; Yoshida *et al.*, 1987; Bovi *et al.*, 1987). Transfection of normal cells with plasmids carrying acidic FGF, (aFGF) cDNA results in the constitutive expression of the growth factor and confer on the cells properties in culture that are generally characteristic of transformed cells (Jaye *et al.*, 1988). Thus, there is reason to believe that bFGF may be involved in the abnormal growth processes characteristic of BPH. We do not know if prostatic bFGF is a stroma inducer responsible for glandular hyperplasia. However, the properties of the molecule and its synthesis by prostatic fibroblasts (Story *et al.*, 1988a) suggests that bFGF could have such a role.

Is bFGF synthesized by the stoma or the epithelial cells of the prostate?

Basic FGF has been purified from many mesoderm-and neuroectoderm-derived tissues (reviewed by Gospodarowicz *et al.*, 1987) including BPH (Story *et al.*, 1987b). The growth factor is also present in the normal prostate. The cell population(s) responsible for bFGF synthesis in the prostate is not known. Studies to identify the prostatic cell(s) responsible for bFGF are in progress. These studies include: isolation of stromal and epithelial enriched cell populations, growth of the cells in culture, and assessment of growth factor production; immunohistochemical staining for bFGF in cultured cells and normal prostate and BPH tissue sections; and Northern analysis and *in situ* hybridization for bFGF mRNA. Our preliminary findings indicate that prostate-derived fibroblasts in culture synthesize bFGF (Story *et al.*, 1988a). Once the cells responsible for bFGF production are identified, studies will be undertaken to identify agents that regulate growth factor synthesis. In addition, it will be important to determine if bFGF synthesis and regulation is the same in normal prostate and BPH.

The endothelial cells of the prostate are another potential source of bFGF. Cultured endothelial cells derived from aorta (Baird and Ling, 1987; Vlodavsky *et al.*, 1987), adrenal cortex (Baird and Ling, 1987; Schweigerer *et al.*, 1987) and cornea (Vlodavsky *et al.*, 1987) synthesize the growth factor. Macrophages are also known to be a source of bFGF (Baird *et al.*, 1985).

Release of bFGF from cells or from extracellular reservoirs

Studies with cultured endothelial cells have provided important insight as to how bFGF may regulate normal growth processes and have suggested mechanisms by which these physiologic processes might be altered in disease states.

Fibroblast growth factors are mitogenic for cultured endothelial cells. The growth factors are known to be angiogenic factors; a property of the molecule demonstrated both *in vitro* and *in vivo* (reviewed by Folkman and Klagsbrun, 1987). Fibroblast growth factors are also produced by cultured endothelial cells. The growth factors have been identified in the conditioned medium but the majority of the growth factors are cell-associated or are found in the extracellular matrix

(ECM). Heparin sulfate, a major glycosaminoglycan (GAG) in the ECM, binds and stabilizes FGFs (Gospodarowicz *et al.*, 1986; Schreiber *et al.*, 1985). The cDNA for bFGF, sequenced form bovine pituitary (Abraham *et al.*, 1986) and from human fibroblasts (Kurokawa *et al.*, 1987), indicates that the growth factor lacks a signal peptide. The mechanism whereby bFGF exits the cell to be deposited in the ECM is unclear. The growth factor may complex with GAG inside the cell and be transported with GAG. Alternatively, there may be as yet unidentified secretory mechanisms that allow the intracellular bFGF to exit without activating cell surface receptors.

At this time it is not clear why bFGF synthesized by cells is not mitogenic for the same cells unless added exogenously. Presumably, bFGF normally bound to heparan-sulfate GAG is not able to interact with cell surface receptors. Heparan-sulfate degradation is a major enzymatic step in the breakdown of the ECM. The degradation of these GAGs, produced experimentally by heparinase and heparitinase treatment (Baird and Ling, 1987), results in decreased adsorption of newly synthesized bFGF to the ECM and releases the growth factor from the ECM. Thus, one of the mechanisms by which tumors, and perhaps some normal cells, induce neovascularization and cell proliferation might be by releasing heparinase-like enzymes. The observations that several tissues, including tumors, are a rich source of heparinase-like enzymes (Vlodavsky *et al.*, 1983; Matzner *et al.*, 1945) support this hypothesis.

In the proposed model for the role of bFGF in BPH, injury (perhaps resulting from infarction, infection, reflux of urine or sexual stimulation) releases bFGF. The growth factor may be released upon cell lyses or by enhanced enzymatic turnover of the ECM. The growth factor stimulates the proliferation of all of the cells involved in the wound healing process. These may include capillary endothelial cells, smooth muscle cells, fibroblasts, and perhaps epithelial cells of the prostate.

What cells in the prostate respond to bFGF?

Membrane bound receptors serve to recognize and combine with stimulatory or inhibitory agents delivered to the cell and translate this recognition function into a biologic action. Fibroblast growth factors interact with a variety of cell types, producing a characteristic set of responses in each cell type (reviewed by Gospodarowicz *et al.*, 1985). By analogy to other polypeptide hormone and growth factor systems, the cellular action of FGF is exerted through its interaction with specific cell surface receptors. To understand the mode of action of FGF, and its role in BPH, it is important to gain information on the functional properties of these receptors. Specifically, it is important to identify the prostatic cell population that is capable of responding to the growth factor and determine if the properties of the receptor in these cells are altered in BPH.

A high affinity membrane receptor for FGF has been identified in BHK cells (Neufeld and Gospodarowicz, 1985), 3T3 cells, mouse skeletal muscle myoblasts (Olwin and Hauschka, 1986), murine and human endothelial cells, fibroblasts (Schreiber *et al.*, 1985), and bovine lens epithelial cells (Moenner *et al.*, 1986). The high affinity FGF receptor typically shows binding affinities (K_D) in the pM range. The reported number of sites per cell varied from 2,000 in murine myoblasts (Moenner *et al.*, 1986) to 120,000 in BHK cells (Neufeld and Gospodarowicz, 1985). The receptor has an apparent M_r of approximately 150,000 (Friesel *et al.*, 1986), but at least two distinct M_r species have been identified (Neufeld and Gospodarowicz, 1985). These receptor properties are similar to those seen for other protein growth factors. In addition, bFGF binds to low affinity, 2 nM K_D, (Moscatelli, 1987) cell-associated heparin-like molecules synthesized by cultured cells and in basement membrane and ECM of animal tissues (Matzner *et al.*, 1985), that complicates attempts to characterize the FGF receptor. To overcome this problem, it will be necessary to perform bFGF receptor studies with isolated membranes derived from fibroblasts and epithelial populations of normal and BPH prostate. As antibodies to the FGF receptor become available they will be of value in this aspect of the study.

What is the distribution of bFGF in the normal prostate and BPH?

Human BPH originates in distinct foci in the periurethral portion of the prostate (McNeal, 1978). The hyperplastic process then expands outward pushing the normal prostate to the periphery. The first lesions of BPH appear in the third and forth decades of life and, thus, precede the clinical symptoms of BPH by 10 to 30 years (McNeal, 1984). Based on McNeal's observations, it seems logical to assume that alterations have occurred in specific cells of the prostate. It is possible that injury occurs to the periurethral ducts releasing bFGF that can then interact with stroma cells surface receptors. If bFGF plays a role in the etiology of BPH, it may be that the growth factor will be elevated in the periurethral zone of the young prostate and in the nodules in progressive BPH. This hypothesis will be tested by quantitating bFGF in small tissue samples from the peripheral, central, and periurethral zones of the young normal prostate and in the central, periurethral and nodules of the hyperplastic gland. In addition, prostate sections will be reacted with antibody to the growth factor and the distribution of bFGF in the same regions determined. As methods for *in situ* hybridization for bFGF mRNA are developed, these techniques will be incorporated into these studies.

Stromal-epithelial interaction in the prostate and the role of androgens

Development of the prostate from the embryonic urogenital sinus and maintenance of the morphology and secretory function of the adult prostate is dependent upon androgens. Castration results in reduction of prostatic weight, cessation of secretory activity and atrophy and death of the prostatic epithelium (Coffey, 1986; Lee, 1981). These events can be restored to the pre-castrated condition by androgen replacement. Insight into possible mechanism involved in androgen induced development and maintenance of prostatic epithelium is found in the work by Cunha and co-workers (reviewed by Cunha, 1984). Their approach involved recombination experiments with urogenital sinus epithelium and mesenchyme from normal, wild-type, mouse embryos and from mouse embryos with testicular feminization syndrome (Tfm). The prostates of mice with Tfm do not develop because of a mutation affecting the androgen receptor. Their studies clearly showed that prostatic development and secretory function only occurred in tissue recombinants prepared with wild-type mesenchyme (androgen-receptor-positive) irrespective of the source of the epithelium, wild-type or Tfm (androgen-receptor negative). They also demonstrate that wild-type urogenital sinus mesenchyme would induce adult urinary bladder from either wild-type or Tfm mice to differentiate into prostate and that the induction was androgen dependent. These studies indicated that mesenchyme and not the epithelium is the target and mediator of androgen-induced prostatic glandular development. These observations are best explained by a hypothesis in which androgen-responsive stromal cells produce a growth factor that elicits epithelial proliferation. It is tempting to speculate that bFGF might play such a role.

There is general agreement that androgens are required for the development of BPH. The observation that BPH does not develop in men castrated prior to puberty supports this conclusion (Moore, 1944). Agreement on whether castration inhibits hyperplasia is by no means universal. Walsh (1984) has reviewed this topic and concluded that castration may be most influential in types of BPH characterized by epithelial outgrowth and least influential in those cases where the pathology is primarily stromal in nature. A recent study by Peters and Walsh (1987) supports this conclusion. They showed that Nafarelin acetate, a potent luteinizing hormone releasing hormone agonist, decreased prostatic volume and improved symptoms of obstruction in BPH. The major morphologic feature resulting from androgen deprivation was regression of the glandular epithelium.

Attempts have been made to determine which cells of human BPH contain androgen receptors. These studies have given conflicting data. Krieg and co-workers (1983) found similar levels of androgen receptor in both epithelium and stroma. Lahtonen and his associates (1983) reported a higher concentration of androgen receptors in the stroma than in the epithelium. Autoradiographic analysis of human glandular BPH showed androgen receptor localization almost exclusively within the epithelial nuclei; whereas, in fibromuscular BPH higher levels of nuclear androgen receptors were found in the stromal cells than in epithelial cells (Peters and Barrack, 1988).

The study by Peters and Walsh (1987) with Naferlin acetate could be explained by a direct effect of androgen on the prostatic epithelium. An alternative explanation is that the stromal cells are stimulated by androgens to produce factors which stimulate epithelial cell growth. In the case of stromal hyperplasia, the stromal cells might produce factors which stimulate their own growth (autocrine regulation) or the epithelium might produce factors which stimulate the stroma (paracrine regulation). The recognition of the prostatic cell population that is responsible for growth factor synthesis and the identification of cells with the capacity to respond to these signals will provide valuable information for choosing between these possibilities.

REFERENCES

Abraham J.A., Mergia A., Whang J.L., Tumola A., Friedman J., Hjerrild K.A., Gospodarowicz D. and Fiddes J.C. 1986. Nucleotide sequence of a bovine clone encoding the angiogenic protein basic fibroblast growth factor. Science 133:545.

Baird A., Bohlen P., Ling N. and Guillemin R. 1985a. Radioimmunoassay for fibroblast growth factor (FGF): release by the bovine anterior pituitary *in vitro*. Regul. Peptides 10:309.

Baird A., Esch F., Bohlen P., Ling N. and Gospodarowicz D. 1985b. Isolation and partial characterization of an endothelial cell growth factor from the bovine kidney: homology with basic fibroblast growth factor. Regul. Peptides 12:201.

Baird A. and Ling N. 1987. Fibroblast growth factors are present in the extracellular matrix produced by endothelial cells *in vitro*: Implications for a role of heparinase-like enzymes in the neovascular response. Biochem. Biophys. Res. Commun. 142:429.

Baird A., Mormed P. and Bohlen P. 1986. Immunoreactive fibroblast growth factor (FGF) in a transplantable chondrosarcoma: Inhibition of tumor growth by antibodies to FGF. J. Cell. Biochem. 30:79.

Birkhoff J.D. 1983. Natural history of benign prostatic hypertrophy. In: "Benign Prostatic Hypertrophy", (F. Hinman, ed.) Springer-Verlag, New York. p.5.

Bovi P.D., Curatola A.M., Kern F.G., Greco A., Ittman M. and Baslico C. 1987 An oncogene isolated by transfection of Kaposi's sarcoma DNA encodes a growth factor that is a member of the FGF family. Cell 50:729.

Coffey D.S. 1986. The biochemistry and physiology of the prostate and seminal vesicles. In: "Campbell's Urology, 5th ed., vol. 1" (P.C. Walsh, R.F. Gittes, A.D. Perlmutter, and T.A. Stamey, eds.) Saunders, Philadelphia. p.283.

Cunha G.R. 1984. Androgenic effects upon prostatic epithelium via trophic influences from stroma. In: "New Approaches to the Study of Benign Prostatic Hyperplasia" (F.A. Kimball, A.F. Buhl, D.B. Carter, eds.) Alan R. Liss, Inc., New York. p.81.

Deming C.L. and Neumann C. 1939. Early phases of prostatic hyperplasia. Surg. Gynec. Obstet. 68:155.

Dickson C. and Peters G. 1987. Potential oncogene product related to growth factors. Nature (London) 326:833.

Doolittle R.F., Hunkapiller M.W., Hood L.E., DeVare S.G., Robbins K.C., Aaronson S.A. and Antoniades H.N. 1983. Simian sarcoma virus oncogene, v-sis, is derived from the gene (oncogenes) encoding a platelet derived growth factor. Science 221:275.

Downward J., Yarden Y., Mayes E., Scrace G., Totty N., Stockwell P., Ullrich A., Schlessinger J. and Waterfield M.D. 1984. Close similarity of epidermal growth factor receptor and v-erb-B oncogene protein sequences. Nature (London) 307:521.

Esch F., Baird A., Ling N., Ueno N., Hill F., Denoroy L., Kleppe R., Gospodarowicz D., Bohlen P. and Guillemin R.L. 1985. Primary structure of bovine pituitary basic fibroblast growth factor (FGF) and comparison with the amino-terminal sequence of bovine brain acidic FGF. Proc. Natl. Acad. Sci. (USA) 82:6507.

Folkman J. and Klagsbrun M. 1987. Angiogenic factors. Science 235:442.

Friesel R., Burges W.H., Mehlman T. and Maciag T. 1986. The characterization of the receptor for endothelial cell growth factor by covalent ligand attachment. J. Biol. Chem. 261:7581.

Gauthier T., Maftouh M. and Picard C. 1987. Rapid enzymatic degradation of (^{125}I) (Tyr 10) FGF (1-10) by serum *in vitro* and involvement in the determination of circulating FGF by RIA. Biochem. Biophys. Res. Commun. 145:775.

Gospodarowicz D. 1974. Localization of a fibroblast growth factor and its effect alone and with hydrocortisone on 3T3 cell growth. Nature (London) 249:123.

Gospodarowicz D. 1985. Biological activity *in vivo* and *in vitro* of pituitary and brain fibroblast growth factor. In: "Mediators in Cell

Growth and Differentiation" (R.J. Ford and A.L. Maizel, eds.) Raven Press, New York. p.109.

Gospodarowicz D., Baird A., Cheng J., Lui G.M., Esch F. and Bohlen P. 1986a. Isolation of fibroblast growth factor from bovine adrenal gland: physiochemical and biological characterization. Endocrinology 118:82.

Gospodarowicz D., Cheng J., Lui G.M., Baird A., Esch F. and Bohlen P. 1985. Corpus luteum angiogenic factor is related to fibroblast growth factor. Endocrinology 117:2383.

Gospodarowicz D., Ferrara N., Schweigerer L. and Neufeld G. 1987. Structural characterization and biological functions of fibroblast growth factor. Endocrine Rev. 8:95.

Gospodarowicz D., Neufeld G. and Schweigerer S. 1986b. Fibroblast growth factor. Mol. Cell. Endocrinol. 46:187.

Halaban R., Ghosh S. and Baird A. 1987. bFGF is the putative natural growth factor from human melanocytes. *In Vitro* Cell Dev. Biol. 23:47.

Hunter J., 1841, Observations on certain parts of the animal economy. In: "The Complete Works of John Hunter, vol. 4", (J.E. Palmer, ed.) Haswell, Barrington, and Haswell, Philadelphia. p.68.

Jacobs S.C. and Lawson R.K. 1980. Mitogenic factors in human prostate extracts. Urology 16:488.

Jacobs S.C. and Pikna D. and Lawson R.K. 1979. Prostatic osteoblastic factor. Invest. Urol. 17:195.

Jaye M., Lyall R.M., Mudd R., Schlessinger J. and Sarver N. 1988. Expression of acidic fibroblast growth factor cDNA confers growth advantage and tumorigenesis to Swiss 3T3 cells. EMBO J. 7:963.

Klagsbrun M., Sasse J., Sullivan R. and Smith J.A. 1986. Human tumor cells synthesize an endothelial cell growth factor that is structurally related to basic fibroblast growth factor. Proc. Natl. Acad. Sci. (USA) 83:2448.

Klagsbrun M., Smith S., Sullivan K., Shing Y., Davidson S., Smith J.A. and Sasse J. 1987. Multiple forms of basic fibroblast growth factor: Amino-terminal cleavages by tumor cell-and brain cell-derived acid proteinases. Proc. Natl. Acad. Sci. (USA) 84:1839.

Krieg M., Bartsch W., Thomsen M. and Voigt K.D. 1983. Androgens and estrogens. Their interactions with stroma and epithelium of human benign prostatic hyperplasia and normal prostate. J. Steroid. Biochem. 19:155.

Kurokawa T., Sasada R., Iwan M. and Igarashi K. 1987. Cloning and expression of cDNA encoding human basic fibroblast growth factor. FEBS Lett. 213:189.

Lahtonen R., Bolton N.J., Kontturi M. and Vihko R. 1983. Nuclear androgen receptors in the epithelium and stroma of human benign prostatic hypertrophic glands. Prostate 4:129.

Lawson R.K., Story M.T., and Jacobs S.C. 1981. A growth factor in extracts of human prostatic tissue. In: "The Prostatic Cell: Structure and Function, Part A", (G.P. Murphy, A.A. Sandberg and J.P. Karr, eds.) Alan R. Liss Inc., New York. p.325.

Lawson R.K. 1986. The natural history of benign prostatic hyperplasia. In: "AUA Update Series, vol. 5" (T.P. Ball, Jr., ed.) American Urological Association, Inc., Houston, Texas. p.2.

Lawson R.K. 1988. Growth factor and benign prostatic hyperplasia. World J. Urol. (in press).

LeDuc I.E. 1939. The anatomy of the prostate and the pathology of early benign hypertrophy. J. Urol. 42:1217.

Lee C. 1981. Physiology of castration-induced regression in rat prostate. In: "The Prostatic Cell: Structure and Function, Part A", (G.P. Murphy, A.A. Sandberg and J.P. Karr, eds.) Alan R. Liss, Inc., New York. p.145.

Lobb R. and Fett J. 1984. Purification of two distinct growth factors from bovine neural tissue by heparin affinity chromatography. Biochemistry 23:6295.

Massoglia S.L., Kenney J.S. and Gospodarowicz D. 1987. Characterization of murine monoclonal antibodies directed against basic fibroblast growth factor. J. Cell Physiol. 132:531.

Matzner Y., Bar-Ner M., Yahalom J., Ishai-Michaeli R., Fuks Z. and Vlodavsky I. 1985. Degradation of heparin sulfate in the subendothelial extracellular matrix by a readily released heparanase from human neutrophile. J. Clin. Invest. 76:1306.

McNeal J.E. 1984. Anatomy of the prostate and morphogenesis of BPH. In: "New Approaches to the Study of Benign Prostatic Hyperplasia, Part A", (F.A. Kimball, A.E. Buhl and D.B. Carter, eds.) Alan R. Liss, Inc., New York. p.145.

Moenner M., Chevallier B., Badet J. and Barritault D. 1986. Evidence and characterization of the receptor to eye-derived growth factor I, the retinal form of basic fibroblast growth factor, on bovine epithelial lens cells. Proc. Natl. Acad. Sci. (USA) 83:5024.

Moore R.A. 1943. Benign hypertrophy of the prostate: a morphological study. J. Urol. 50:680.

Moore R.A. 1944. Benign hypertrophy and carcinoma of the prostate: occurrence and experimental production in animals. Surgery 16:152.

Moscatelli D. 1987. High and low affinity binding sites for basic fibroblast growth factor on cultured cells; absence of a role for low affinity binding in the stimulation of plasminogen activator production by bovine capillary endothelial cells. J. Cell Physiol. 131:123.

Moscatelli D., Presta M., Joseph-Silverstein J. and Rifkin D.B. 1986. Both normal and tumor cell produce basic fibroblast growth factor. J. Cell Physiol. 129:273.

Neufeld G. and Gospodarowicz D. 1985. The identification and partial characterization of the fibroblast growth factor receptor of baby hamster kidney cells. J. Biol. Chem. 260:13860.

Neufeld G. and Gospodarowicz D. 1986. Basic and acidic fibroblast growth factors interact with the same cell surface receptors. J. Biol. Chem. 261:5631.

Olwin B.B. and Hauschka S.D. 1986. Identification of the fibroblast growth factor receptor of Swiss 3T3 cells and mouse skeletal muscle myoblasts. Biochemistry 25:3487.

Peters C.A. and Barrack E.R. 1986. Androgen receptor localization in the prostate using a new method of steroid receptor autoradiography. In: "Benign Prostatic Hyperplasia, vol. 2", (C.H. Rogers, ed.) U.S. Govt. Printing Office, Washington, D.C. p.175.

Peters C.A. and Walsh P.C. 1987. The effect of nafarelin acetate, a luteinizing-hormone-releasing hormone agonist, on benign prostatic hyperplasia. New England J. Med. 317:599.

Reischauer F. 1925. Die entystehung der sogenannten prostatahypertrophie. Virchow's Arch. Path. Anat. 256:357.

Schreiber A.B., Kenney J., Kowalski W.J., Friesel R., Mehlman T. and Maciag T. 1985. Interaction of endothelial cell growth factor with heparin: characterization by receptor and antibody recognition. Proc. Natl. Acad. Sci. (USA) 82:6138.

Schweigerer L., Neufeld G., Friedman J., Abraham J.A., Fiddes J.C. and Gospodarowicz D. 1987. Capillary endothelial cells express basic fibroblast growth factor, a mitogen that promotes their own growth. Nature 325:257.

Shing Y., Folkman J., Sullivan R., Butterfield C., Murray J. and Klagsbrun M. 1984. Heparin affinity: purification of a tumor-derived capillary endothelial cell growth factor. Science 223:1296.

Story M.T., Esch F., Shimasaki S., Sasse J., Jacobs S.C. and Lawson R.K. 1987a. Amino-terminal sequence of a large form of basic fibroblast growth factor isolated from human benign prostatic hyperplastic tissue. Biochem. Biophys. Res. Commun. 142:702.

Story M.T., Jacobs S.C. and Lawson R.K. 1983. Epidermal growth factor is not the major growth-promoting agent in extracts of prostatic tissue. J. Urol. 30:175.

Story M.T., Jacobs S.C. and Lawson R.K. 1984a. Partial purification of a prostatic growth factor. J. Urol. 132:1212.

Story M.T., Jacobs S.C. and Lawson R.K. 1984b. Preliminary characterization and evaluation of techniques for the isolation of prostate-derived growth factor. In: "New Approaches to the Study of Benign Prostatic Hyperplasia", (F.A. Kimball, A.E. Buhl, D.B. Carter, eds.) Alan R. Liss, Inc., New York. p.197.

Story M.T., Jacobs S.C. and Lawson R.K. 1988a. Identification of basic fibroblast growth factor in cultured fibroblasts from normal and BPH prostate. J. Urol. 139:190A.

Story M.T., Sasse J., Jacobs S.C. and Lawson R.K. 1987b. Prostatic growth factor: Purification and structural relationship to basic fibroblast growth factor. Biochemistry 26:3843.

Story M.T., Sasse J., Kakuska D., Jacobs S.C. and Lawson R.K. 1988b. A growth factor in bovine and human testes that is structurally related to basic fibroblast growth factor. J. Urol. 140:422.

Ueon N., Baird A., Esch F., Ling N. and Guillemin R. 1987. Isolation and partial characterization of basic fibroblast growth factor from bovine testes. Mol. Cell. Endocrinol. 49:189.

Vlodavsky I., Folkman J., Sullivan R., Fridman R., Ishni-Michaeli R., Sasse J. and Klagsbrun M. 1987. Endothelial cell-derived basic fibroblast growth factor: synthesis and deposition into subendothelial extracellular matrix. Proc. Natl. Acad. Sci. (USA) 84:2292.

Vlodavsky I., Fuks Z., Bar-Ner M., Ariav Y. and Schirrmacher V. 1983. Lymphoma cells mediated degradation of sulfated protoglycans in the subendothelial extracellular matrix: relationship to tumor cell metastasis. Cancer Research 43:2704.

Walsh P.C. 1984. Human prostatic hyperplasia: etiological considerations. In: "New Approaches to the Study of Benign Prostatic Hyperplasia", (F.A. Kimball, A.E. Buhl and D.B. Carter, eds.) Alan R. Liss, Inc., New York. p.1.

Yoshida T., Miyagawa K., Odagiri H., Sakamoto H., Little PFR., Terada M. and Sugimura T. 1987. Genomic sequence of hst, a transforming gene encoding a protein homologous to fibroblast growth factors and the int-2-encoded protein. Proc. Natl. Acad. Sci. (USA) 84:7305.

MORPHOGENESIS AND CYTODIFFERENTIATION OF MALE SEX ACCESSORY EPITHELIA: INVOLVEMENT OF THE MESENCHYME AND NEUROTRANSMITTERS

Stephen J. Higgins*
Gerald R. Cunha[x]
Peter Young[+]
Annemarie A. Donjacour[+] and
E. Margaret Kinghorn*

*Department of Biochemistry
University of Leeds
Leeds, LS2 9JT United Kingdom

[+]Department of Anatomy
and
[x]Center for Reproductive Endocrinology
University of California Medical Center
San Francisco, CA. 94143

Androgenic steroid hormones characteristically regulate the expression of specific genes within their target organs (Higgins and Parker, 1980; Cunha *et al.*, 1987). Since the advent of recombinant DNA technology, considerable progress has been made in our understanding of how androgens and other steroid hormones selectively activate gene expression (Evans, 1988; Green and Chambon, 1988). With the recent cloning of the androgen receptor (AR) gene (Lubahn *et al.*, 1988; Chang *et al.*, 1988 a,b; Trapman *et al.*, 1988), further dramatic advances can be expected. Little attention, however, has been given to the possible roles of paracrine and autocrine mechanisms in androgen action. This chapter describes some of our studies concerning such mechanisms in the functional cytodifferention of male sex accessory organs and in the maintenance of the differentiated state.

ANDROGENIC REGULATION OF SEMINAL VESICLE SECRETORY PROTEIN SYNTHESIS

The model system used in these studies is based on the seminal vesicles (SV) of the male rodent. In these animals the paired SV are among the most prominent of the accessory organs of the reproductive tract. Normal SV morphology and histology are particularly characteristic, the most notable features being the folded epithelium composed of tall columnar cells (Price and Williams-Ashman, 1961; Williams-Ashman, 1983). These cells are highly specialized for secretion, possessing an extensive rough endoplasmic reticulum, a prominent Golgi complex and numerous apically-located secretory vesicles (Szirmai and van der Linde, 1965; Dahl *et al.*, 1973; Brandes, 1974). The SV lumen is normally filled with an intensely eosinophilic secretion.

The bulk of the protein in rat SV secretion is accounted for by SV secretory (SVS) proteins I-V (Higgins *et al.*, 1976; Ostrowski *et al.*, 1979; Fawell *et al.*, 1986). These are the structural proteins of the copulatory plug (Higgins *et al.*, 1976; Fawell *et al.*, 1986; Fawell and Higgins, 1987). Their synthesis is highly androgen-dependent, decreasing dramatically after testosterone is removed following castration or on administration of an anti-androgen (Higgins *et al.*, 1976; Ostrowski *et al.*, 1979, 1982). Normal rates of synthesis are restored by androgen replacement therapy. Immunocytochemistry of SV tissue sections with polyclonal antibodies monospecific for the individual rat SVS proteins has shown that the synthesis of SVS protein is confined to the SV epithelium (SVE; Fawell and Higgins, 1986). Furthermore, the SVE appears to be functionally homogeneous in that every epithelial cell synthesizes all of the major SVS proteins (Fawell and Higgins, 1986).

The effects of androgens on SVS protein synthesis have been particularly well studied with respect to proteins IV and V. Their genes have been cloned and sequenced (Mansson *et al.*, 1981; McDonald *et al.*, 1983; Williams *et al.*, 1983; Kandala *et al.*, 1983). Nucleotide probes have been used to show that androgens control the synthesis of proteins IV and V principally via changes in the amounts of their specific messenger RNAs (Higgins and Burchell, 1978; Mansson *et al.*, 1981; Kandala *et al.*, 1983; Fawell and Higgins, 1984). 'Run on' transcription studies in nuclei isolated from the SV of rats of different androgen status have shown that androgen specifically stimulates transcription of genes IV and V, though major post-transcriptional effects are undoubtedly involved as well (Hemingway and Higgins, in preparation).

MODEL FOR ANDROGEN ACTION

Androgen regulation of specific gene expression in SVE conforms to the generally accepted model for steroid hormone action (see Higgins and Gehring, 1978; Rousseau, 1984). This model, formulated from observations on a wide variety of steroid responsive systems, envisages that the steroid hormone binds with high affinity to intracellular protein receptors present only in target cells. The receptor steroid complex acts as a transcriptional regulatory factor by binding to steroid response elements, cis-acting genetic sequences with the properties of enhancers, associated with the responsive genes (Hatzopoulos *et al.*, 1988). While the response elements have been characterized for many genes, including the androgen response elements in the long terminal repeat sequences of mouse mammary tumor virus (Ham *et al.*, 1988), those for SVS protein genes IV and V have yet to be identified and described.

Although the model outlined above adequately describes many aspects of androgen action in SV, a number of important aspects are overlooked. These include the participation of stromal cells with which the epithelial cells form a close association and which are known to possess androgen receptors (Cooke, 1988) and are thus potential targets for androgens. In the regulation of SVS protein synthesis it is also assumed that the only extracellular inducer is the androgen, with no participation of any other endocrine, paracrine or autocrine factors.

THE STROMA AS AN INDUCER OF EPITHELIAL DEVELOPMENT

There is a large body of evidence that inductive effects from the mesenchyme, the presumptive stroma, are essential for epithelial morphogenesis (Cunha *et al.*, 1983, 1987; Kedinger, 1986; Haffen *et al.*, 1987; Kratochwil, 1987; Sakakura, 1987). In the case of androgen-dependent accessory organs this has been shown by the use of heterotypic tissue recombinants in which the nature of the mesenchyme determines the morphogenetic fate of the epithelium with which it is associated (Cunha *et al.*, 1983, 1987). Some of these recombinants contained tissues from mice carrying the Tfm mutation. Such tissues are normally insensitive to androgens because they fail to express functional androgen receptors (Gehring *et al.*, 1971; Attardi and Ohno, 1974; Lubahn *et al.*, 1988). However, in combination with a genetically normal (XY) mesenchyme, Tfm epithelia appeared to undergo normal morphogenetic development (Cunha and Lung, 1978; Cunha and Chung, 1981; Shannon and Cunha, 1984). These results raise the possibility that androgen receptors are not an obligatory requirement for hormonal induction of epithelial gene expression.

INVOLVEMENT OF NEUROTRANSMITTERS IN ANDROGEN ACTION

In the ventral prostate of the rat, androgens induce the synthesis of prostatic steroid binding protein or prostatein (Heyns and DeMoor, 1977; Parker *et al.*, 1978; Lea *et al.*, 1979). Chung and his group have presented evidence that prostatein expression involves neurotransmitters such as noradrenaline and dopamine (Thompson *et al.*, 1987). The possibility presents itself, therefore, that neurotransmitters may be among the induction factors through which the stroma regulates epithelial morphogenesis and function.

DIRECTIVE INFLUENCE OF THE MESENCHYME ON FUNCTIONAL DEVELOPMENT OF SEMINAL VESICLE EPITHELIUM

The paired foetal Wolffian ducts (WD) develop regionally along morphologically and functionally distinct lines. The upper (cranial) portion gives rise to the epididymis, the ductus deferens develops from the middle region and the SV buds off from the lower (caudal) portion. Thus, the WD provides a group of embryologically related tissues in which the role of mesenchyme in determining epithelial development may be conveniently investigated.

Morphogenetic and functional development of upper Wolffian duct grafts

Before exploring whether SV mesenchyme (SVM) can instructively induce SV morphogenesis and functional cytodifferentiation in WD epithelia (WDE), we first confirmed that the upper portion of the WD does indeed give rise to epididymal tissue. Specimens of this region of the WD from foetal male rats (16-17 days of gestation) were grafted into the kidney capsule of adult male hosts where the grafts become vascularized and exposed to circulating androgens. After 3-4 weeks, the grafts were recovered and analyzed (Table 1). As expected, the histology was uniformly indicative of epididymal tissue with no signs of any tissue resembling SV. Antibodies specific for two secretory proteins B/C and D/E, normally expressed in mature epididymis (Brooks and Higgins, 1980; Brooks and Tiver, 1983; Brooks *et al.*, 1986), gave strong immunocyto-chemical staining of the epithelium in these grafts and both proteins were detected by immunoblotting extracts of the grafts. No evidence could be obtained for any proteins resembling SVS proteins.

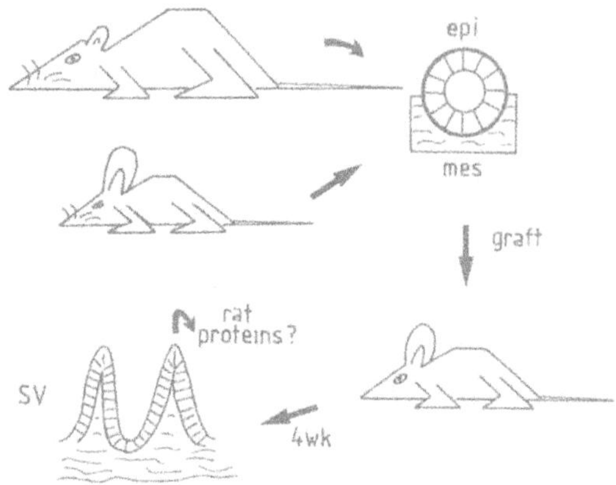

Figure 1. Construction and growth of tissue recombinants. Upper WD from foetal male rats (Table I) were treated with trypsin at 4°C and the epithelium (WDE) was separated by microdissection. Seminal vesicles from neonatal male mice (0-24h *post partum*) were likewise treated to provide mesenchyme (SVM). Upper WDE and SVM were combined *in vitro* and then grafted in renal capsules of adult male hosts (Table I). After 4 weeks for growth the recombinants were recovered, the secretion was prepared for SDS-PAGE and tissues were fixed in p-formaldehyde, paraffin wax embedded and sectioned for histology and immunocyto-chemistry.

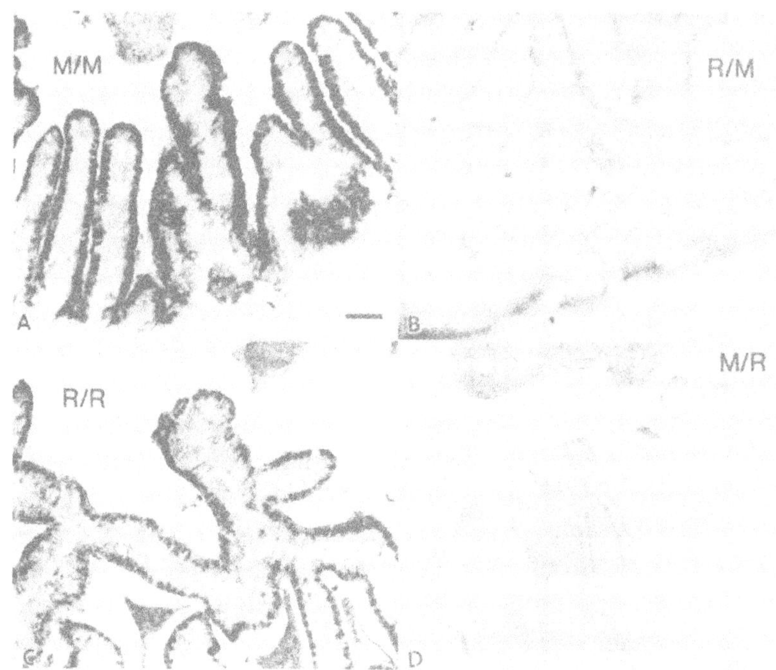

Figure 2. Species specificity of antibody probes. Rodent SV tissue sections were used for immunocytochemistry. The primary antibody was visualized by a peroxidase-linked second antibody system. Panel A, normal mouse SV with anti-mouse SVS antibody (M/M). Bar 25um (same for all panels). Panel B, normal mouse SV with anti-rat SVS protein IV (R/M). Panel C, normal rat SV with anti-rat SVS protein IV (R/R). Panel D, normal rat SV with anti-mouse SVS (M/R). Note positive epithelial staining in panels A and C, negative staining in panels B and D.

Table 1. Functional development of rat upper Wolffian duct rudiments.

Morphology	Epididymis
Epididymal protein B/C	Present
Epididymal protein D/E	Present
SVS protein IV	Not detected
SVS protein V	Not detected

The cranial portion (approximately the upper third) was removed from the Wolffian ducts of fetal (16-17th day of gestation) male rats and grown as subcapsular renal grafts (21 total) in adult male athymic ('nude') mice. After 3-4 weeks the grafts were removed and examined histologically (hematoxylin and eosin staining). The indicated tissue-specific proteins were assayed by immunocytochemistry and by immunoblotting of secretory proteins separated by SDS-PAGE. Details of the methods will be found in Fawell and Higgins (1987) Fawell *et al.*, (1987) and Higgins *et al.*, (1989a,b). Epididymal antibodies were a gift from Dr. D. E. Brooks.

Instructive induction of WDE by SVM

Figure 1 shows the way in which heterologous tissue recombinants were constructed using upper WDE and SVM. After 4 weeks of growth in adult male hosts, the tissue recombinants were examined for evidence of SV epithelial morphogenesis and functional cytodifferentiation. However, any SV tissue detected in these recombinants could have arisen from traces of SVE incompletely removed from the SVM rather than from instructive induction of the WDE. Although this is unlikely since specimens of SVM grafted alone showed no epithelial morphogenesis (Higgins *et al.*, 1989a), we used heterospecific (rat/mouse) tissue recombinants (see Fig. 1) to discount this possibility in a definitive manner. True instructive induction should result in rat SVS proteins being produced, whereas tissue contamination would be indicated by the presence of mouse SVS proteins. Immunological methods may be used to distinguish between the SVS proteins of the two species (Higgins *et al.*, 1989a) as shown in Figure 2. Anti-rat SVS protein IV gave a strong immunocytochemical staining of rat SVE but not of mouse SVE, while an antibody against mouse SVS proteins gave the opposite result.

The tissue obtained from rat upper WDE + mouse SVM recombinants (shown in Figure 3) had the characteristic morphology of mature SV with a highly convoluted secretory epithelium composed of tall columnar cells. Intensely eosinophilic secretion filled the luminal spaces. Immunocytochemistry with antibodies against rat SVS proteins IV and V resulted in strong staining of the apical aspects of all the epithelial cells in every section examined. The luminal secretion was also heavily stained. In contrast anti-mouse SVS antibody failed to stain the sections. Thus the SVE present in these recombinants must have arisen by instructive induction of the WDE by the SVM.

The secretion of the recombinants contained all five rat SVS proteins as shown by SDS-PAGE (Figure 4, lane 40 and this was confirmed by immunoblotting samples of the secretion with antibodies against each of the individual SVS proteins (Higgins *et al.*, 1989b). Reciprocal tissue recombinants, i.e. mouse upper WDE + rat SVM, also resulted in instructive induction of SVE, but here the SVE synthesized SVS proteins characteristic of mouse and not rat SV (Figure 4, lane 3). Once again the identities of all the mouse SVS proteins were confirmed by

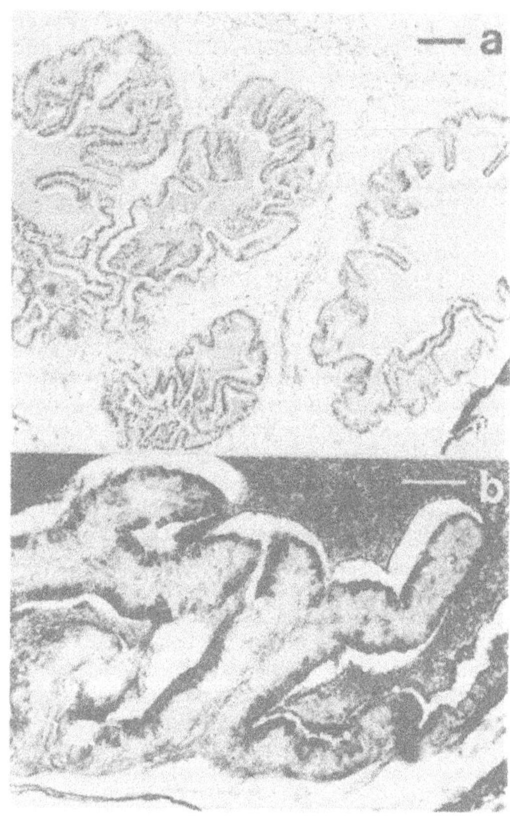

Figure 3. Instructive induction of SVE in heterotypic recombinants. Rat upper WDE was combined with mouse SVM and grown as renal grafts as described in Figure 1. Panel a, hematoxylin and eosin; note SV morphology (bar, 100um). Panel b, immunocytochemistry with anti-rat SVS protein IV; note positive staining of epithelium and secretion (bar, 25 μm). Staining with anti-mouse SVS was negative (not shown).

Immunoblotting (Higgins *et al.*, 1989b). In these examples of instructive induction of SVE, the functional reprogramming was complete in that there was no evidence for the synthesis of the androgen-dependent epididymal marker proteins B/C or D/E (Higgins *et al.*, 1989b).

The epithelium of the middle region of the WD, the progenitor of the ductus deferens, was also instructively induced by SVM to form SVE. As before the SVE produced SVS proteins appropriate to the species (rat or mouse) that had provided the middle WDE of the recombinant (Figure 4, lanes 1 and 2).

ARE ANDROGEN RECEPTORS REQUIRED FOR FUNCTIONAL CYTODIFFERENTIATION OF SEMINAL VESICLE EPITHELIUM?

Development of SVE in Tfm tissue recombinants

The experimental protocol adopted for these investigations followed that shown in Figure 1 except that WDE from foetal TfmY mice was combined with SVM from normal XY rats. However, the male offspring of the breeding pairs used in these experiments may include both normal (XY) and Tfm (TfmY) genotypic animals. While these may be readily distinguished phenotypically when mature, they are indistinguishable in the embryo. In order to ascertain retrospectively the genotype of the embryonic tissue used for each tissue recombinant, the urogenital sinus (UGS) from each embryo was grafted alongside the tissue recombinant. If the UGS came from an XY fetus, it would be able to respond to the host's androgens and would therefore develop into prostatic tissue. Conversely, if the UGS was from a TfmY fetus it would be unable to respond to androgens and conse-quently would develop as vaginal tissue. Only tissue recombinants con-structed with the WDE from fetuses whose UGS was clearly Tfm in character were examined further. While most of the recombinants containing TfmY epithelium failed to show development, a small proportion (2 of 14) did produce tissues that had all the histological characteristics of normal SV (Fig. 5). To confirm that the SVE in these latter recombinants was of mouse origin, the tissue was stained with Hoechst dye 33258 (Cunha and Vanderslide, 1984; Higgins *et al.*, 1989a). With this dye the nuclei of mouse tissues typically show an intense but irregular fluorescent staining with many fluorescent foci due to satellite DNA. In contrast, the nuclei of rat tissues are usually less intensely stained and are generally devoid of fluorescent foci. A shown in Figure 5, the SVE of the TfmY + XY recombinants clearly displayed a fluorescent staining pattern indicative of mouse tissue, whereas the stroma was of rat origin.

Functional cytodifferentiation in Tfm epithelium

Is the SVE present in these Tfm recombinants functionally normal? The copious secretion present in the luminal spaces of the recombinants (Fig. 5) was certainly suggestive of normal SV secretory activity, so we explored this aspect of SVE development by immunocytochemical staining (Fig. 5). The apical portions of the epithelial cells were readily stained with antibody against mouse SVS protein, as too was the luminal secretion. None of the cells were stained with anti-rat SVS protein IV

We stress the preliminary nature of the data described here. We do not know whether SV epithelial function in the Tfm recombinants was entirely normal, especially as regards the complete range of mouse SVS (Fig. 5). Thus, a TfmY SVE appears to be able to synthesize androgen-dependent proteins diagnostic of XY SVE provided that the TfmY epithelium is associated with an XY mesenchyme.

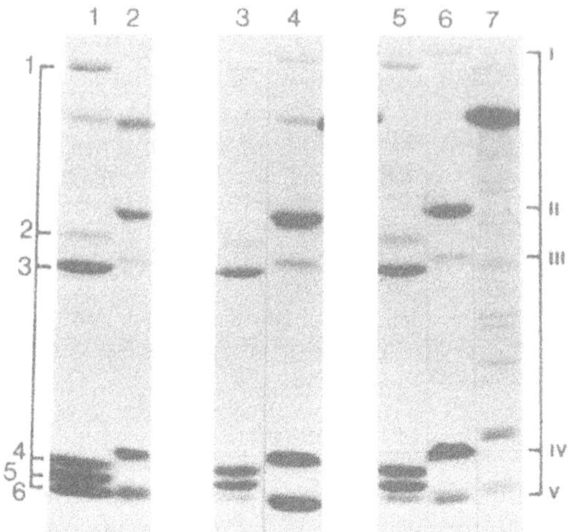

Figure 4. Electrophoretic analysis of proteins secreted by WD tissue recombinants. Secretory proteins (20-50 μg per lane) from recombinants of WDE + SVM were separated by SDS-PAGE and stained with Coomassie blue. Lane 1, mouse middle WDE + rat SVM; lane 2, rat middle WDE + mouse SVM; lane 3, mouse upper WDE + rat SVM; lane 4, rat upper WDE + mouse SVM; lane 5, mouse SV secretion; lane 6, rat SV secretion; lane 7, rat epididymal secretion.

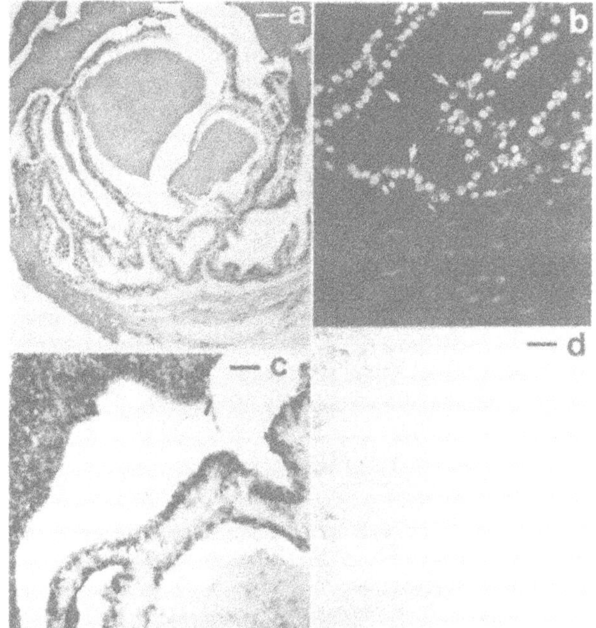

Figure 5. Development of Tfm tissue recombinants. Tissue recombinants composed of WDE from foetal TfmY mice (14-15th day of gestation) and SVM from neonatal rats were constructed and grown as indicated in Figure 1. Panel a, hematoxylin and eosin; note SV morphology. Panel b, Hoechst dye 33258; note fluorescent foci in epithelial nuclei (arrows). Panel c, anti-mouse SVS; note staining of epithelium and secretion. Panel d, anti-rat SVS protein IV; note background staining. Bars, 100 μm (panel a) and 25 μm (panels b-d).

proteins expressed or the quantitative aspects of their expression. Moreover, we have not yet confirmed that the Tfm SVE remained devoid of androgen receptors by use of either [^{3}H]-dihydrotestosterone autoradiography (Shannon and Cunha, 1984) or more recently available antibody and nucleotide probes (Lubahn *et al.*, 1988). The proportion of Tfm recombinants developing as renal grafts was low (2 out of 14) compared with 55% of control recombinants (Xy WDE + XY SVM) (Higgins *et al.*, 1989b). The reasons for this are unclear, but development might be much more exquisitely and crucially dependent on the exact gestational age of the tissue or the precise physical juxtapositioning of the mesenchyme and epithelium than in recombinations between normal tissues. However, the fact that some tissue recombinants containing Tfm WDE did differentiate as SV demonstrates the ability of this epithelium to respond via interactions with an AR^+ stroma. More importantly, our confidence in the correctness of our interpretations of these Tfm SV recombinants is considerably strengthened by the results of parallel experiments in which a Tfm epithelium was induced to undergo prostatic cytodifferentiation.

Development of prostate in Tfm tissue recombinants

A similar type of experiment has been performed using bladder epithelium (BLE) from adult TfmY mice and urogenital sinus mesenchyme (UGM) from XY mouse or XY rat fetuses (Donjacour *et al.*, 1988). In this case there is no uncertainty concerning the genotype of the BLE, since the tissue is taken from an adult mouse with the Tfm phenotype, i.e. small undecended testis, lack of the male secondary sex organs, small phallus, no uterus or oviducts and a small blind-ended vagina (Lyons and Hawkes, 1970). The rat UGM + Tfm BLE recombinants, when transplanted into male host mice, gave rise to prostatic tissue at approximately the same frequency as did rat UGM + mouse XY BLE recombinants. Prostatic differentiation was judged by morphology (hematoxylin and eosin staining) as well as by immunocytochemistry with antibodies to androgen-dependent prostatic secretory proteins (Donjacour *et al.*, 1988). The species origin of the epithelium was ascertained by Hoechst dye staining (Donjacour *et al.*, 1988). These data are also preliminary since only a small number of tissue recombinants have been analyzed in this manner. However, prostatic morphology has been observed in numerous XY UGM + Tfm BLE recombinants (Cunha and Lung, 1978; Cunha and Chung, 1981; Shannon and Cunha, 1984), and in a limited number of these specimens the Tfm epithelium has expressed androgen-dependent prostatic secretory proteins (Donjacour *et al.*, 1988).

The conclusion that is strongly suggested by all the data presented in this section is that epithelial androgen receptors are not necessary for SV or prostatic morphogenesis. In addition, it is likely that androgen-dependent epithelial secretory activity can be initiated and maintained without epithelial receptors. Mesenchymal androgen receptors appear to be the mediators of androgen-dependent effects in both the mesenchyme and the epithelium.

INVOLVEMENT OF NEUROTRANSMITTERS IN ANDROGEN ACTION

The results of the experiments described in the previous section raise fundamental questions regarding the mechanism of action of androgens in SVE. In an attempt to define more precisely the requirements for androgen regulated gene expression in SVE, we investigated the conditions under which SV epithelial cells could be induced to express SVS proteins *in vitro* (Kinghorn *et al.*, 1987).

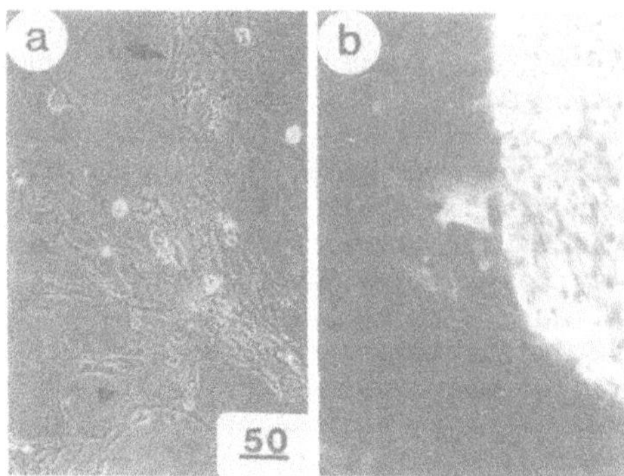

Figure 6. Growth of SV cells in primary culture. Cells from enzymatically disrupted SV tissue of castrated rats were grown for 5 days on collagen-coated coverslips in medium containing serum plus testosterone and 5α-dihydrotestosterone (each 10^{-8}M). Panel a: phase contrast micrograph showing an epithelial cell mass (arrow) and adjacent fibroblasts (arrow-head). Panel b: immunocytochemistry with a monoclonal antibody against human cytokeratin visualized with a rhodamine-conjugated second anti-body. Note the positive fluorescence of the epithelial cells and background fluorescence of fibroblasts. Magnification in (μm) applies to both panels. Adapted from Kinghorn *et al.*, (1987) with permission.

Adult rats were castrated in order to repress the endogenous synthesis of SVS proteins. After 2-4 weeks, the SV were removed and enzymically disrupted. The SV cells were then plated on collagen-coated surfaces and grown in Dulbecco's modified Eagle's medium containing 10% fetal calf serum. Under these conditions small clumps of cells readily attached to the collagen substratum and provided foci for epithelial cells to grow out and form sheets. Figure 6 shows that the cells in such a sheet expressed cytokeratin, an epithelial cell marker. Other cell types with a fibroblast-like appearance were also present in these cultures, but these did not express cytokeratin (Fig. 6). They are presumed to be of stromal origin. A detailed description of these primary cell cultures will be found in Kinghorn *et al.* (1987).

Although the epithelial cells in these cultures grew well in the serum containing medium, neither testosterone or 5α-dihydrotestosterone (DHT) at concentrations of 10^{-8}M was able to induce the synthesis of SVS proteins IV or V. Various attempts were made to modify the culture conditions so that androgen induction could be achieved. Use of collagen gels, feeder cell layers, and extracellular matrices were all ineffective. Supplementation of the medium with other hormones (hydrocortisone, corticosterone, insulin and prolactin), growth factors (epidermal growth factor, cholera toxin), cyclic nucleotides (cyclic AMP, cyclic GMP, dibutyryl cyclic AMP) or a variety of other factors also failed to result in androgenic induction of SVS proteins.

However, Chung and his group reported that there is a link between the levels of certain neurotransmitters in prostatic tissue and the expression of the androgen responsive prostate marker protein, prostatein (Thompson *et al.*, 1978). With this in mind, we grew the SV cultures for 3 days in medium containing testosterone and DHT supplemented with acetylcholine, noradrenaline and serotonin (each at 10^{-8}M). Under these conditions, the epithelial cells did express SVS proteins IV and V. Figure 7 shows the immunocytochemical staining of the epithelial cells

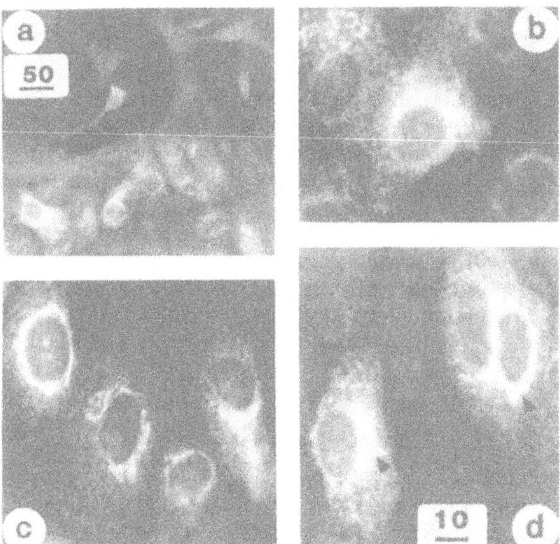

Figure 7. Neurotransmitters are required for androgenic induction of SV proteins. Seminal vesicle cells were grown in culture in the presence of androgens (see Fig. 6) together with acetylcholine, noradrenaline and serotonin (each 10^{-8}M). Cells were screened with anti-SVS protein IV and a rhodamine-conjugated second antibody (panels a-d). In panel a, note the positive immunofluorescence of epithelial cells but not the fibroblasts (arrows). Magnifications are in μm (panels b-d at same magnification). Adapted from Kinghorn *et al.*, (1987) with permission.

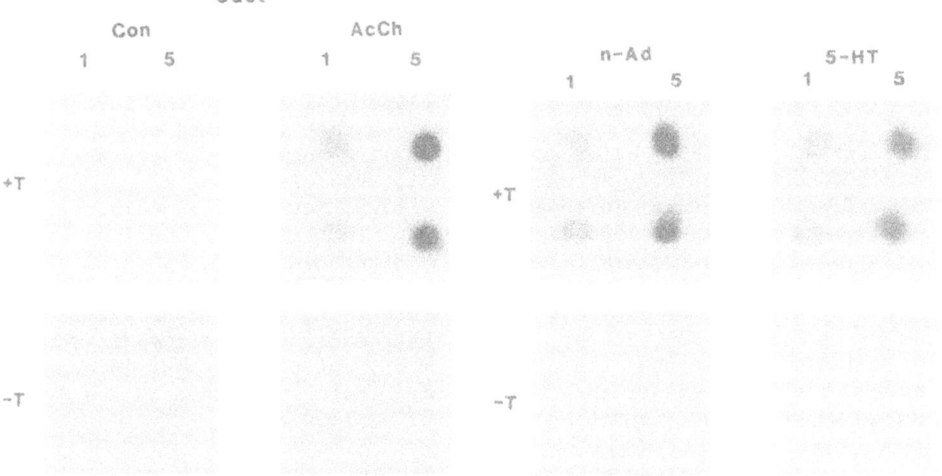

Figure 8. Neurotransmitters and androgens induce secretion of SV proteins. Seminal vesicle cells were grown in serum containing medium plus acetylcholine (AcCh), noradrenaline (n-Ad) or serotonin (5-HT) each 10^{-8}M) or in the absence of neurotransmitters (Con). In each case testosterone and DHT (each 10^{-8}M) were either added (+T) or omitted (-T) from the medium. After 3 days the medium was removed and the medium proteins were isolated by acetone precipitation, solubilized in SDS and spotted onto nitrocellulose strips in 1 μg (1) or 5 μg (5) duplicate samples. The strips were screened with anti-rat SVS protein IV followed by [^{125}I]Protein A and then autoradiographed. Adapted from Kinghorn *et al.*, (1987) with permission.

with anti-rat SVS protein IV. Fibroblastic stromal cells were not stained.

The immunocytochemical data were reinforced by immunoblotting the proteins extracted from the culture medium (Figure 8). Androgenic induction of SVS protein IV was only seen when one of the neurotransmitters was also present. Any one of the neurotransmitters was effective; there were no additive or synergistic effects with all three neurotransmitters. Since none of the neurotransmitters induced SVS proteins on its own in the absence of androgens, we conclude that the neurotransmitters act as permissive agents for androgens rather than as inducers in their own right.

RECONSIDERATION OF THE MECHANISM OF ANDROGEN ACTION IN THE SEMINAL VESICLES

Stromal-epithelial tissue interactions are important in many aspects of development and in the normal functioning of mature organ systems (Cunha *et al.*, 1983, 1985, 1987; Haffen *et al.*, 1986; Kedinger *et al.*, 1987; Kratochwil, 1987; Sakakura, 1987). It is not surprising, therefore, that androgen-dependent synthesis of SVS proteins by SVE requires the participation of stromal cells. What was unexpected in the studies discussed above is that in combination with a normal stroma, epithelia from animals whose tissues are normally thought of as androgen insensitive due to the absence of functional androgen receptors (Gehring *et al.*, 1971; Attardi and Ohno, 1974; Lubahn *et al.*, 1988) can be induced to express genes characteristic of androgen responsive epithelia. Failure to detect androgen receptors in the Tfm epithelium of Xy UGM + Tfm BLE recombinants using steroid autoradiography (Shannon and Cunha, 1984) appears to rule out the mesenchymal induction of epithelial receptors as an explanation for the epithelial response.

The biochemical mechanism of stromal-epithelial tissue interactions is still unclear. Depending on the developmental system, the inductive process may be brought about by cell-cell contact (Lehtonen *et al.*, 1975; Thesleff *et al.*, 1977; Meier and Hay, 1975), changes in the extracellular matrix (Grobstein, 1967; Bernfield *et al.*, 1984; Perris *et al.*, 1988) or via diffusible factors (Toivonen *et al.*, 1976; Karkinen-Jaaskelainen, 1978; Grunz and Tacke, 1986; Smith, 1987). The latter may include well-established growth factors (Kimelman and Kirschner, 1987; Slack *et al.*, 1987; Rosa *et al.*, 1988), many of which are known to be present in accessory glands, particularly the prostate (Schuurmans *et al.*, 1988; Maddy and Habib, 1988; Storey, 1989). What mechanism operates in the SV is entirely unknown, but the involvement of neurotransmitters in androgen action *in vitro* suggests that the inducing factors produced by the SV stroma might include neurotransmitters of various kinds.

Seminal vesicles do possess characteristic cholinergic and adrenergic innervations, but these are chiefly associated with the smooth musculature and vasculature (Setchell and Brooks, 1988). Several neurotransmitters have been detected in the SV of various species (Setchell and Brooks, 1988). Evidence for direct innervation of the secretory mucosa or the involvement of neurotransmitters in the secretory process in vivo is very sparse. In the dog, stimulation of the hypogastric nerve or administration of systemic cholinomimetic drugs indices copious prostatic secretion. In other species, however nervous regulation of secretory activity of accessory gland epithelia, or generally in androgen action, is unclear. Neurotransmitters appear to be involved in the androgenic regulation of prostatein synthesis in rat ventral prostate

(Thompson *et al.*, 1987). Androgens regulate β-adrenergic receptors in the same organ (Collins *et al.*, 1988).

In fact, our own experimental results do not permit us to conclude that neurotransmitters act directly on SV epithelial cells. The permissive effect of neurotransmitters on androgen action in SV cell cultures could be exerted indirectly via the stromal cells. In this case, the neurotransmitters, instead of themselves being the inducing factors for the epithelium, might collaborate with androgens in stimulating stromal cells to produce those factors.

The results described here raise fundamental questions regarding our view of how androgenic steroids regulate specific gene expression, in particular concerning the role of androgen receptors in the process. In addition, they link the nervous and endocrine systems and provide some of the first evidence for paracrine effects in androgen action. Obviously much more work is needed before a clear and complete picture will emerge.

ACKNOWLEDGEMENTS

We thank Joel Brody for skilled technical assistance, John Warrallo for photography and Paula Duncan for typing the manuscript. The work was undertaken during the tenure (by SJH) of an American Cancer Society-Eleanor Roosevelt-International Cancer Fellowship awarded by the International Union Against Cancer. Financial support was also provided by the Medical Research Council, U.K. (Project Grant G83/046/58CA) and the Yorkshire Cancer Research Campaign, U.K. (to SJH), and by NIH grants HD 21919 and HD 17491 (to GRC).

REFERENCES

Attardi B. and Ohno S. 1974. Cytosol androgen receptor from kidney of normal and testicular ferminized (Tfm) mice. Cell 2:205.

Bernfield M.R., Banerjeem S.D., Koda, J.E. and Raprager A.C. 1984. Remodelling of the basement membrane as a mechanism of morphogenetic tissue interaction. In: "Role of Extracellular Matrix in Development", (R.L. Trestad, ed.), Alan R. Liss, Inc., New York, p. 545.

Brandes D. 1974. Fine structure and cytochemistry of male accessory organs. In: "Male Accessory Sex Organs: Structure and Function". (D. Brandes, ed.), Academic Press, New York, p. 18.

Brooks D.E. and Higgins S.J. 1980. Characterisation and androgen-dependence of proteins associated with luminal fluid and spermatozoa in the rat epididymis. J. Reprod. Fertil. 59:363.

Brooks D.E. and Tiver K. 1983. Localisation of epididymal secretory proteins on rat spermatozoa. J. Reprod. Fertil. 69:651.

Brooks D.E., Means A.R., Wright E.S., Singh S.P. and Tiver K.K. 1986. Molecular cloning of cDNA for androgen-dependent sperm-coating glycoproteins secreted by rat epididymis. Eur. J. Biochem. 161:13.

Chang C., Kokontis J. and Liao S. 1988a. Molecular cloning of human and rat complementary DNA encoding androgen receptors. Science 240:324.

Chang C., Kokontis J. and Liao S. 1986b. Structural analysis of complementary DNA and amino acid sequences of human and rat androgen receptors. Proc. Natl. Acad. Sci. USA 85:7211.

Collins S., Quarmby V.E., French F.S., Lefkowitz R.J. and Caron M.G. 1988. Regulation of the β-2-adrenergic receptor and its mRNA in the rat ventral prostate by testosterone. FEBS Lett. 233:173.

Cooke P.S. 1988. Ontogeny of androgen receptors in male mouse reproductive organs. Endocrinology 122:92.

Cunha G.R. and Chung L.W.K. 1981. Stromal-epithelial interactions. I. Induction of prostatic phenotype in urothelium of testicular feminized (Tfm/Y) mice. J. Steroid Biochem. 14:1317.

Cunha G.R. and Lung B. 1978. The possible influence of temporal factors in androgenic responsiveness of urogenital tissue recombinants from wild-type and androgen-insensitive (Tfm) mice. J. Exp. Zool. 205:181.

Cunha G.R. and Vanderslice K.D. 1984. Identification in histological sections of species origin of cells from mouse, rat and human. Stain Technology 59:7.

Cunha G.R., Chung L.W.K., Shannon J.M., Taguchi O. and Fuji H. 1983. Hormone-induced morphogenesis and growth: role of mesenchymal-epithelial interactions. Recent Prog. Horm. Res. 39:559.

Cunha G.R., Bigsby R.M., Cooke P.S. and Sugimura Y. 1985. Stromal-epithelial interactions in adult organs. Cell Differentiation 17:137.

Cunha G.R., Donjacour A.A., Cooke P.S., Mee S., Bigsby R.M., Higgins S.J. and Sugimura Y. 1987. Endocrinology and developmental biology of the prostate. Endocrine Rev. 8:338.

Dahl E., Kjaerheim A. and Tveter K.J. 1973. The ultrastructure of the accessory sex organs of the male rat. I. Normal structure. Z. Zellforsch. 137:345.

Donjacour A.A. and Cunha G.R. 1989. Seminal vesicle mesenchyme induces prostatic morphology and secretion in urinary bladder epithelium. J. Cell Biol. In press.

Evans R.M. 1988. The steroid and thyroid hormone receptor superfamily. Science 240:889.

Fawell S.E. and Higgins S.J. 1984. Androgen regulation of specific mRNAs, endoplasmic reticulum and Golgi system. Mol. Cell. Endocrinol. 37:15.

Fawell S.E. and Higgins S.J. 1986. Tissue distribution, developmental profile and hormonal regulation of androgen-responsive secretory proteins of rat seminal vesicle studied by immunocytochemistry. Mol. Cell. Endocrinol. 48:39.

Fawell S.E. and Higgins S.J. 1987. Formation of rat copulatory plug: purified seminal vesicle secretory proteins serve as transglutaminase substrates. Mol. Cell. Endocrinol. 53:149.

Fawell S.E., Pappin D.J.C., McDonald C.J. and Higgins S.J. 1986. Androgen-regulated proteins of rat seminal vesicle secretion constitute a structurally-related family present in the copulatory plug. Mol. Cell. Endocrinol. 45:205.

Fawell S.E., McDonald C.J. and Higgins S.J. 1987. Comparison of seminal vesicle secretory proteins of rodents using antibody and nucleotide probes. Mol. Cell. Endocrinol. 50:107.

Gehring U., Tomkins G.M. and Ohno S. 1971. Effect on the androgen-insensitivity mutation on a cytoplasmic receptor for dihydrotesterone. Nature New Biol. 232:106.

Green S. and Chambon P. 1988. Nuclear receptors enhance our understanding of transcription regulation. Trends in Genetics 4:309.

Grobstein C. 1967. Mechanisms of organogenic tissue interaction. Nat. Cancer Inst. Monogr. 26:279.

Grunz H. and Tacke L. 1986. The inducing capacity of the presumptive endoderm of _Xenopus laevis_ studied by transfilter experiments. Wilhelm Roux' Arch. Devel. Biol. 195:467.

Haffen K., Kedinger M. and Simon-Assmann P. 1987. Mesenchyme-dependent differentiation of epithelial progenitor cells in the gut. J. Ped. Gast. Nutrit. 6:14.

Ham J., Thompson A., Needham M., Webb P. and Parker M. 1988. Characterisation of response elements for androgens, glucocorticoids and progestins in mouse mammary tumour virus. Nucleic Acids Res. 16:5263.

Hatzopoulos A.K., Schlokat U. and Gross P. 1988. Enhancers and other _cis_-acting regulatory sequences. _In_: "Transcription and Splicing". (B.D. Hames and D.M. Glover, eds.), IRL Press, Oxford, p. 43.

Heyns W. and DeMoor P. 1977. Prostatic binding protein. A steroid-binding protein secreted by rat prostate. Eur. J. Biochem. 78:221.
Higgins S.J. and Burchell J.M. 1978. Effects of testosterone on messenger RNA and protein synthesis in rat seminal vesicle. Biochem. J. 174:543.

Higgins S.J. and Gehring U. 1978. Molecular mechanisms of steroid hormone action. Adv. Cancer Res. 28:313.

Higgins S.J., Burchell J.M. and Mainwaring W.I.P. 1976. Androgen-dependent synthesis of basic secretory proteins by rat seminal vesicle. Biochem. J. 158:271.

Higgins S.J., Young P., Brody J.R. and Cunha G.R. 1989a. Induction of functional cytodifferentiation in the epithelium of tissue recombinants. I. Homotypic seminal vesicle recombinants. Development. In Press.

Higgins S.J., Young P. and Cunha G.R. 1989b. Induction of functional cytodifferentiation in epithelium of tissue recombinants. II. Instructive induction of Wolffian duct epithelia by neonatal seminal vesicle mesenchyme. Development. In press.

Kandala J.S., Kistler M.K., Lawther R.P. and Kistler W.S. 1983. Characterisation of a genomic clone for rat seminal vesicle secretory protein IV. Nucleic Acids Res. 11:3169.

Karkinen-Jaaskelainen M. 1978. Transfilter lens induction in avian embryo. Differentiation 12:31.

Kedinger M., Haffen K. and Simon-Assman P. 1986. Control mechanisms in the ontogenesis of villus cells. In: "Molecular and Cellular Basis of Digestion", (P. Desnuelle, H. Sjostrom, and O. Noren, eds.), Elsevier, Amsterdam, p. 323.

Kimelman D. and Kirschner M. 1987. Synergistic induction of mesoderm by FGF and TGF-β and the identification of an mRNA coding for FGF in the early Xenopus embryo. Cell 51:869.

Kinghorn E.M., Bate A.S. and Higgins S.J. 1987. Growth of rat seminal vesicle epithelial cells in culture: neurotransmitters are required for androgen-regulated synthesis of tissue-specific secretory proteins. Endocrinology 121:1678.

Kratochwil K. 1987. Tissue combination and organ culture studies in the development of the embryonic mammary gland. In: "Developmental Biology: A Comprehensive Synthesis (R.B.L. Gwatkin, ed.), Plenum Press, New York, p. 315.

Lea O.A., Petrusz P. and French F.S. 1979. Prostatein. A major secretory protein of the rat ventral prostate. J. Biol. Chem. 254:6196.

Lehtonen E., Wartiovaara J., Nordling S. and Saxen L. 1975. Demonstration of cytoplasmic processes in Millipore filters permitting kidney tubule induction. J. Embryol. Exp. Morph. 33:187.

Lubahn D.B., Joseph D.R., Sullivan P.M., Willard H.F., French F.S. and Wilson E.M. 1988. Cloning of human androgen receptor complementary DNA and localisation to the X chromosome. Science 240:327.

McDonald C., Williams L., McTurk P., McIntosh E. and Higgins S. 1983. Isolation and characterisation of genes for androgen-responsive secretory proteins of rat seminal vesicles. Nucleic Acids Res. 11:917.

Maddy S.Q. and Habib F.K. 1988. EGF receptors: a depleted expression in cancer of the prostate. J. Endocrinol. 117:243.

Mansson P.E., Sugino A. and Harris S.E. 1981. Use of a cloned double-stranded cDNA coding for a major androgen dependent protein in rat seminal vesicle secretion: the effect of testosterone in gene expression. Nucleic Acids Res. 9:935.

Meier S. and Hay E.D. 1975. Stimulation of corneal differentiation by interaction between cell surface and extracellular matrix. J. Cell Biol. 66:275.

Ostrowski M.C., Kistler M.K. and Kistler W.S. 1979. Purification and cell-free synthesis of a major protein from rat seminal vesicle secretion. J. Biol. Chem. 254:383.

Ostrowski M.C., Kistler M.K. and Kistler W.S. 1982. Effect of castration on synthesis of seminal vesicle secretory protein IV in the rat. Biochemistry 21:3525.

Parker M.G., Scrace G.T. and Mainwaring W.I.P. 1978. Testosterone regulates the synthesis of major proteins in rat ventral prostate. Biochem. J. 170:115.

Perris R., von Boxberg Y. and Lofberg J. 1988. Local embryonic matrices determine region-specific phenotypes in neural crest cells. Science 241:86.

Price D. and Williams-Ashman H.G. 1961. The accessory reproductive glands of mammals. In: "Sex and Internal Secretion", (W.C. Young, ed.), Williams and Wilkins, Baltimore, p. 366.

Ross F., Roberts A.B., Danielpour D., Dart L.L., Sporn M.B. and David. I.B. 1988. Mesoderm induction in amphibians: The role of TGF-β-2-like factors. Science 239:783.

Rousseau G.G. 1984. Control of gene expression by glucocorticoid hormones. Biochem. J. 224:1.

Sakakura T. 1987. Mammary embryogenesis. In: "The Mammary Gland: Development Regulation and Function", (M.C. Neville and C.W. Daniel, eds.), Plenum Press, New York, p. 37.

Schuurmans A.L.G., Bolt J. and Mulder E. 1988. Androgens and transforming growth in factor β modulate the growth response to epidermal growth factor in human prostatic tumour cells (LNCaP). Mol. Cell. Endocrinol. 60:101.

Setchell B.P. and Brooks D.E. 1988. Anatomy, vasculature, innervation and fluids of the male reproductive tract. In: "The Physiology of reproduction", (E. Knobil and J. Neill, eds.), Raven Press, New York. p. 753.

Shannon J.M. and Cunha G.R. 1984. Characterisation of androgen binding and DNA synthesis in prostate-like structures induced in the urothelium of testicular feminized (Tfm/Y) mice. Biol. Reprod. 31:175.

Slack J.M.W., Darlington B.G., Heath J.K. and Godsave S.F. 1987. Mesoderm induction in early *Xenopus* embryos by heparin-binding growth factors. Nature (London) 326:197.

Smith J.C. 1987. A mesoderm-inducing factor is produced by a *Xenopus* cell line. Development 99:3.

Story M.T. 1989. Prostate growth factor: characterisation and its role in normal prostate and benign prostatic hyperplasia. In: "Paracrine and Autocrine Interactions in Reproductive Endocrinology", (L.C. Krey, B.J. Gulyan, and J.M. McCracken, eds.), Plenum Press, New York. p.

Szirmai J.A. and van der Linde P.C. 1965. Effect of castration on the endoplasmic reticulum of the seminal vesicle and other target epithelia in the rat. J. Ultrastruct. Res. 12:380.

Thompson T.C., Zhau H. and Chung L.W.K. 1987. Catecholamines are involved in the growth and expression of prostatic binding protein by rat ventral prostatic tissues. Prog. Clin. Biol. Res. 239:239.

Toivonen S., Tarin D. and Saxen L. 1976. The transmission of morphogenic signals from amphibian mesoderm to ectoderm in primary induction. Differentiation 5:49.

Trapman J., Klaassen P., Kuiper G.G.J.M., van der Korput J.A.G.M., Faber P.W., van Rooij H.C.J., Geurts van Kessel A., Voohorst M.M., Mulder E. and Brinkmann A.O. 1988. Cloning, structure and expression of a cDNA encoding the human androgen receptor. Biochem. Biophys. Res. Commun. 153:241.

Williams L., McDonald C., Jackson S., McIntosh E. and Higgins S. 1983. Isolation and characterisation of genomic and cDNA clones for an androgen-regulated secretory protein of rat seminal vesicle. Nucleic Acids Res. 11:5921.

Williams-Ashman H.G. 1983. Regulatory features of seminal vesicle development and function. Current Topics in Cell Regulation 22:201.

THE ROLE OF STROMAL-EPITHELIAL INTERACTIONS IN THE REGULATION OF GROWTH AND DIFFERENTIATION IN ADULT EPITHELIAL CELLS

Gerald R. Cunha*
Stephen J. Higgins+
Annemarie A. Donjacour*
Norio Hayashi* and
Peter Young*

*Department of Anatomy and
Reproductive Endocrinology Center
University of California
San Francisco, CA 94143

+Department of Biochemistry
University of Leeds
Leeds, United Kingdom

There is general agreement that paracrine influences are of paramount importance during embryonic and fetal life. From studies on amphibians it has been shown that the differentiation of mesodermal tissue results from an influence of cells of the vegetal pole upon cells of the animal pole (Nieuwkopp, 1973). Once formed, the mesoderm is involved in inducing neural development and in specifying the cranial to caudal pattern on the embryonic axes such that head structures form cranially and tail structures caudally. During subsequent organogenesis, a series of secondary inductions between mesenchyme and epithelium are involved in the morphogenesis, growth and functional cytodifferentiation of a variety of organ systems (Saxen *et al.*, 1980; Ekblom, 1984; Haffen *et al.*, 1987; Kedinger *et al.*, 1986; Sawyer *et al.*, 1983; Saunders and Gasseling, 1968; Kratochwil, 1987; Kollar, 1972, 1987).

Organogenetic processes are initiated during embryonic or fetal periods. Many organs attain functional activity before birth as a prerequisite for impending postnatal life (liver, pancreas, lungs, kidneys). Other organs, particularly those of the reproductive system, are rudimentary at birth; these organs complete their morphogenesis and express functional activity during postnatal periods. Of course for all organs, morphogenetic processes continue to lay down new architectural units from birth until final mature size is attained. It is reasonable that postnatal morphogenesis proceeds by fundamentally the same developmental mechanisms that are operative during earlier prenatal periods. However, the question arises as to the exact role of stromal-epithelial interactions in adulthood after morphological and functional maturity is attained.

For all adult organs, functional activity and epithelial morphology must be maintained as epithelial cells undergo senescence, die, and are replaced. Cellular replacement may be rapid, as in the case of the intestine, vagina, and skin (Leblond and Cheng, 1976; Evans and Long, 1920; Potten, 1983; Potten and Hendry, 1983) or exceedingly slow, as in the case of the urinary bladder (Hicks, 1975). Clearly, cellular proliferation must be regulated within bounds compatible with maintenance of normal structure and function. In systems containing undifferentiated stem cell populations, such as vagina, skin, or the intestine, an orderly process of epithelial differentiation and functional maturation must be maintained as cells continually die an dare replaced. Thus, in both the fetus and the adult it is quite likely that differentiation processes are basically similar. In addition, for the adult male and female genital tracts and mammary gland true morphogenetic processes occur cyclically in both male and female seasonal breeders, as well as in females during estrous and menstrual cycles, pregnancy, and lactation. During reproductive cycles, pre-existing parenchymal and stromal elements may be degraded and later regenerated from rudimentary anlagen. Morphogenetic periods may be followed in turn by periods of functional activity. These true morphogenetic events in adult reproductive organs closely resemble the primary developmental events that occur in the perinatal period.

In this review, we will specifically focus upon the role of stromal-epithelial interactions during postnatal periods particularly in adulthood. In this regard the following issues will be addressed. Are adult epithelial cells responsive to inductive influences form connective tissue? Do adult stromal cells themselves have the ability to induce epithelium? What is the role of stroma in androgen-induced epithelial growth? What is the relevance of stromal-epithelial interactions to proliferative disorders such as carcinogenesis?

RESPONSIVENESS OF POSTNATAL EPITHELIAL CELLS TO INDUCTIVE INFLUENCE FROM STROMA

Barring neoplastic transformation, adult epithelial cells generally exhibit a remarkable phenotypic stability with normal differentiation continuing to be maintained throughout life. This stability of differentiation could be due to an irreversible determination of the epithelial cells themselves or, alternatively, to ongoing instructive messages from stroma or the extracellular matrix. The fact that various isolated epithelial cells such as hepatocytes, mammary epithelial cells, and keratinocytes continue to maintain their differentiated properties under certain *in vitro* conditions argues for an intrinsic irreversible determination of epithelial differentiation (Bissell *et al.*, 1987; Lee *et al.*, 1984; Blum *et al.*, 1987; Reid and Jefferson, 1984; Michalopoulos and Pitot, 1975; Fusenig, 1986; Fusenig *et al.*, 1981; Lillie *et al.*, 1983). However, analysis of tissue recombinants in which adult epithelial cells are grown in association with inductive mesenchyme clearly demonstrates that a variety of postnatal epithelia can be induced to undergo remarkable changes in cytodifferentiation.

One of the more striking examples of the developmental plasticity of an adult epithelium is the induction of prostatic differentiation in adult urinary bladder epithelium (BLE) by mesenchyme of the urogenital sinus (UGM). The induced BLE forms a secretory epithelium that resembles prostate as assessed by histological, histochemical, and ultrastructural criteria. Moreover, the induced BLE expressed androgen receptors,prostate specific antigens and exhibits a dependence upon androgens for proliferation. Proteins synthesized by UGM + adult BLE recombinants resemble those of the prostate based upon two-dimensional gel electrophoretic analysis (Cunha *et al.*, 1980a; 1983a; Neubauer *et al.*, 1983). Similarly, vaginal epithelium from 20-day-old mice is induced to express the prostatic phenotype when grown in association with UGM (Cunha, 1975).

Since UGM is the inducer of fetal prostate, its effect on adult prostatic epithelium has also been examined (Norman *et al.*, 1986). Small ≥300 micron fragments of adult mouse prostatic ducts containing 250 to 1500 epithelial cells were associated with UGM from 16-day-old embryonic rats. Under these conditions, the mouse prostatic epithelial cells proliferated rapidly, and during 1 month of growth in male hosts, the tissue recombinants formed 30 to 40 milligrams (wet weight) of prostatic tissue. This means that adult prostatic epithelial cells were induced to undergo ductal morphogenesis and increased their numbers approximately 16,000-fold in one month; that is, the original 1500 mouse prostatic cells gave rise to approximately 24 million prostatic epithelial cells during this period of growth (Normal *et al.*, 1986). Grafts of small fragments of adult prostatic ducts are maintained by themselves but fail to grow. Recent studies utilizing antisera reactive to secretory proteins specific to individual prostatic lobes have demonstrated that, when fragments of ventral prostatic ducts are induced to proliferate in response to UGM, the resultant prostatic tissue expresses secretory proteins unique to the dorso-lateral prostate (Hayashi, Donjacour and Cunha, unpublished). Thus, UGM has morphogenetic, proliferative and differentiative effects on adult prostatic epithelial cells as is also the case for UGM + BLE recombinants. Comparable studies have been described for the mammary gland (Daniel and DeOme, 1965; Daniel *et al.*, 1965, 1968; Daniel and Young 1971; Sakakura *et al.*, 1979a).

Even more striking are the results of tissue recombinants composed of seminal vesicle mesenchyme (SVM) and epithelium of either the ureter (UrE) or ductus deferens (DDE) of adult rats and mice. For this study, tissue recombinants were prepared heterospecifically, eg., rat mesenchyme + mouse epithelium or vice versa. In tissue recombinants composed of rat SVM + 60-day-old mouse UrE or mouse SVM + 60-day-old rat UrE, the adult UrE differentiated into SV epithelium whose origin corresponded to the species from which the UrE was derived based upon staining with Hoechst dye 33258 (Cunha and Vanderslice, 1984). These results were confirmed by using species specific immunologic probes to rat and mouse SV secretory proteins (Higgins *et al.*, 1986). In tissue recombinants composed of rat SVM + adult mouse UrE, the apical portion of the induced columnar epithelial cells exhibited intense staining with antibodies specific to mouse but not rat SV secretory proteins. In contrast, tissue recombinants composed of mouse SVM + adult rat UrE formed SV tissue that expressed rat, but not mouse SV secretory proteins (Fig. 1a-d). Immunocytochemical findings were confirmed by SDS-PAGE (Fig. 1e) and Western blot analysis. Identical results were obtained for tissue recombinants composed of rat SVM + adult mouse DDE and mouse SVM + adult rat DDE (Cunha *et al.*, 1988). In addition to the changes in adult epithelial function cytodifferentiation elicited in epithelia of the ureter or ductus deferens, a great deal of epithelial proliferation had occurred, resulting in a substantial increase in the size of the tissue recombinants. A striking increase in tissue recombinant size was also seen in tissue recombinants composed of SVM associated with small fragments of adult SV mucosa. In this case, the SV phenotype was maintained, but the adult SVE was induced to proliferate forming large masses of functionally active SV tissue (Higgins *et al.*, 1989a). Thus, adult epithelial cells from several sources (prostate, seminal vesicle, mammary gland, bladder, ureter, and ductus deferens) are responsive to inductively active mesenchyme which can stimulate epithelial proliferation, alter adult epithelial cytodifferentiation, and may even elicit expression of a completely new functional phenotype.

These recent findings on the responsiveness of adult epithelia to mesenchymal influences extend earlier studies on the adult epidermis that demonstrated that the dermis induces and maintains regional patterns of adult epithelial differentiation. Epidermal thickness and patterns of keratinization vary regionally such that the epidermis exhibits different thicknesses and patterns of keratinization on the sole of the foot, tail, ear, trunk, gingiva, hard palate and alveolar mucosa. In a large series of experiments in which connective tissue from one region was associated with epithelium from another region, the characteristics of the adult epidermal cells generally changed in accord with the source of the connective tissue with which the epithelium was experimentally associated (Billingham and Silvers, 1968; Spearman, 1974; Karring *et al.*, 1971; Bernimoulin and Schroeder, 1980; Briggaman, 1982; Botero *et al.*, 1975; Donn, 1978). While the changes elicited by heterotypic stroma in adult epidermal cells are readily apparent, they merely represented minor modulations in epithelial differentiation since the basic stratified squamous phenotype was maintained in all cases. However, implicit in all of the above examples in which adult stromal cells elicited changes in epithelial thickness and keratinization is the fact that adult stroma had inductive activity, as discussed below.

INDUCTIVE ACTIVITIES OF ADULT STROMA CELLS

Adult stromal cells can function as true instructive inductors capable of eliciting major changes in epithelial morphology, growth, cytodifferentiation, and even functional activity. Although few adult systems are amenable to such analysis, it has been possible to demonstrate that adult vaginal stoma can induce vaginal differentiation in neonatal uterine epithelial cells (Cunha, 1976). In these experiments the normally simple, columnar uterine epithelium (UTE) was induced by adult vaginal stroma (VS) to form a keratinized, stratified squamous epithelium. This change in epithelial phenotype does not appear to be a simple metaplasia because, when adult VS + UE recombinants were grafted into intact female hosts, the epithelium cycled from a mucified to a cornified state through the estrous cycle in concert with the host's vaginal epithelium (Cunha, 1976). This change in epithelial

Figure 1. Immunocytochemistry of heterospecific tissue recombinations of seminal vesicle mesenchyme + adult ureteral epithelium (SVM + UrE). Rat SVM + mouse UrE (A,B) and mouse SVM + rat UrE (C,D) grown in male hosts for 4 weeks. Note the complexly folded mucosa characteristic of seminal vesicle in A-D. Anti-mouse SV secretion produces positive staining (immunoperoxidase) in the epithelium of tissue recombinants composed of (A) rat SVM + mouse UrE, but not in (B) mouse SVM + rat UrE. Conversely, anti-rat SVS-V produces positive staining in (C) mouse SVM + rat UrE, but not in (D) rat SVM + mouse UrE. Note specific staining of apical portions of cells in insets. Original magnification = 80x, A-D; insets 320, 500x. (E) SDS-PAGE of SV secretions (SVS) of both mice and rats as well as secretory products from heterospecific recombinations of mouse SVM + rat UrE and rat SVM + mouse UrE. Rat SVS contains the 5 major proteins (I-V), while mouse SVS contains the 6 major proteins (1-6). The secretions extracted from tissue recombinants whose epithelium was rat in origin contained the 5 major proteins characteristic of rat SVS while the secretory product from the recombinants whose epithelium was mouse in origin has the 6 major proteins characteristic of the mouse. These findings indicate that the adult UrE of mice and rats were induced to change their biochemical expression and to produce SV specific secretory proteins.

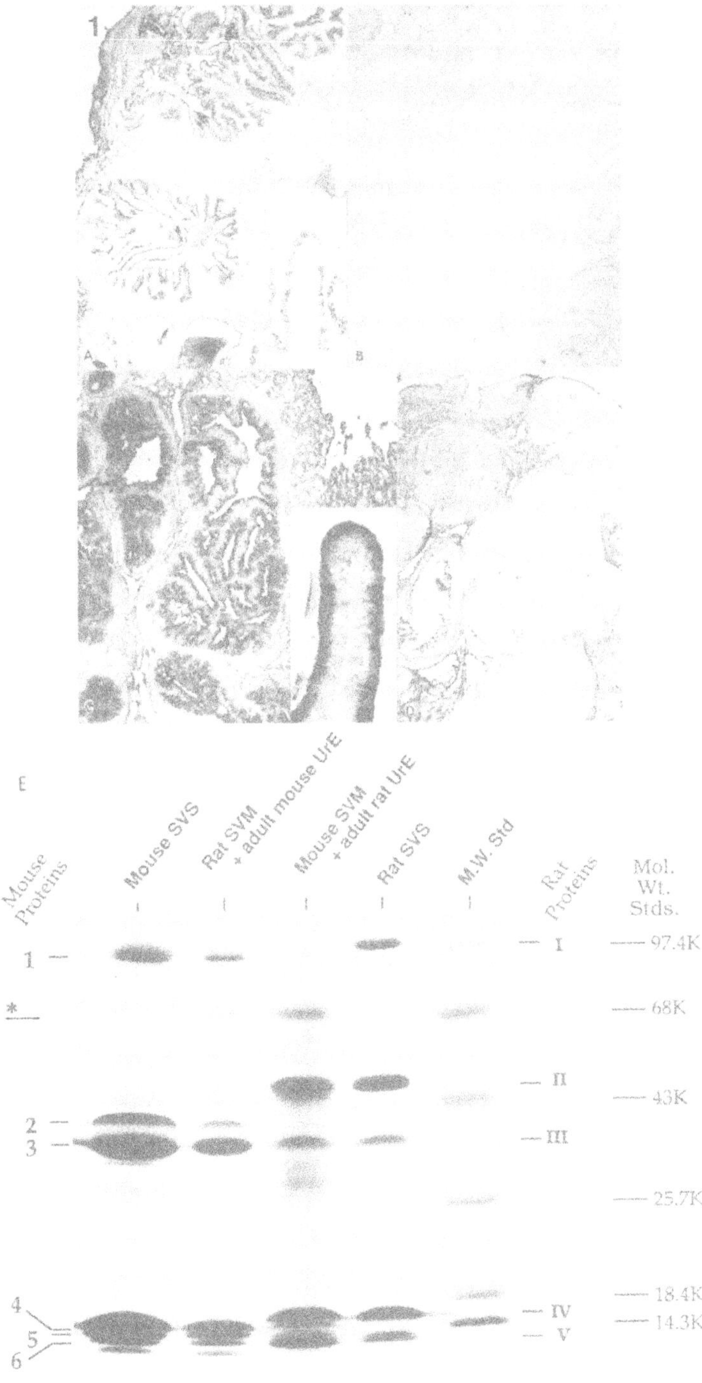

*Albumin contamination

differentiation during the estrous cycle is one of the unique functional hallmarks of vaginal epithelium. Thus, the induction of vaginal differentiation by adult vaginal stroma involves fundamental changes in uterine epithelial differentiation at both the morphological and functional levels. Neonatal uterine and vaginal stroma cultured *in vitro* for 1 to 2 months is also capable of instructively and permissively inducing uterine and vaginal differentiation, respectively, in responsive epithelia (Cooke *et al.*, 1987).

Connective tissues of a variety of integumental derivatives are inductive as discussed above, and in the mammary gland the adult fat pad has the ability to support morphogenesis and growth of both embryonic and adult mammary epithelial cells (Daniel *et al.*, 1965; Sakakura *et al.*, 1979b). Epidermal appendages such as feathers and hairs are induced to form and grow in adulthood by cells of the dermal papillae (Lillie and Wang, 1943; Wang, 1943; Cohen, 1965, 1969; Olifer, 1968; Ibrahim and Wright, 1977; Jahoda *et al.*, 1984). Undoubtedly, regenerative processes in adulthood following wounding are due in part to ongoing epithelial-stromal interactions in which the stroma influences epithelial proliferation, morphogenesis and differentiation.

THE ROLE OF EPITHELIAL-STROMAL INTERACTIONS IN PROSTATIC GROWTH

While androgens are essential for prostatic growth, cell-cell interactions are also of critical importance, an idea underscored by the following experiment. When prostatic ducts obtained from a sexually mature intact adult male mouse were grafted beneath the renal capsule of an intact male host, the graft was maintained but did not grow (Normal *et al.*, 1986). However, when a fetal urogenital sinus (prostatic anlagen) or a neonatal prostatic rudiment was grafted as above, each formed an entire prostate (Cunha, 1972, 1975; Chung and Cunha, 1983; Cunha, unpublished). Thus, prostatic ducts from intact adult males to not grow in intact male hosts, whereas embryonic or neonatal prostatic anlage are capable of considerable growth under identical conditions. This paradox may be explained on basis of epithelial-stromal interactions.

When prostatic development begins in the fetus, small epithelial prostatic buds invade the large mass of mesenchyme surrounding the endodermal urogenital sinus. At this stage the epithelial-mesenchymal ratio greatly favors the mesenchyme, and growth potential is at its highest since the small embryonic rudiment has the potential to form a complete adult organ. As prostatic development proceeds, the epithelial ducts elongate and arborize within the mesenchyme (Sugumura *et al.*, 1986a) and the epithelial-mesenchymal ratio eventually changes to favor the epithelium, i.e. the epithelial:stromal ratio ultimately becomes about 5:1 in adulthood in the rat (DeKlerk and Coffey, 1978).

Studies show that experimental alteration of the epithelial-mesenchymal ratio in the developing prostate has profound effects on prostatic growth and emphasize the importance of this parameter in determining final organ size. Chung and Cunha (1983) dissociated urogenital sinuses (prostatic rudiments) into epithelium and mesenchyme (UGE and UGM). Homotypic UGM + UGE recombinants were prepared in which the amount of UGM was held constant at 1X (the amount obtained from a 16-day embryonic urogenital sinus), while the amount of UGE was varied from 1X to 0.01X. Conversely, another group of tissue recombinants was prepared in which the UGM was varied from 0.1X to 2X keeping the UGE constant at 1X. All recombinants were grown for 1 month in intact male hosts, and then wet weight and DNA content were determined. The results showed that changing the amount of UGE combined with a constant (1X) amount of UGM did not influence the final prostatic mass attained by the recombinants (Chung and Cunha, 1983). Conversely, when the amount of UGE was held constant (1X) and the amount of UGM varied, final tissue mass increased in proportion to the amount of UGM used. These findings suggest that the amount of mesenchyme determines final organ size. This concept receives additional support from studies of tissue interactions between adult prostatic ducts and UGM. Segments of prostatic ducts (PR) containing 200-1500 epithelial cells from intact adult mice were grafted either by themselves or in combination with UGM or bladder mesenchyme (BLM). Grafts of individual ducts were maintained but did not grow when grafted by themselves or when recombined with BLM (Normal *et al.*, 1986). However, UGM + PR recombinants produced 30 to 40 mg (wet weight) of prostatic tissue containing hundreds of prostatic ductal growth in tissue recombinants composed of a prostatic ductal tip with 1X to 8X UGM indicates that new prostatic tissue is formed in proportion to the amount of UGM utilized (Neubauer *et al.*, 1986). Also consistent with these findings are the studies of Chung's group (Chung *et al.*, 1984; Thompson and Chung, 1986) which have shown an increase in prostatic growth *in situ* when UGM or an intact urogenital sinus was grafted directly into the prostate of intact mature males. All of the above observations are consistent with the idea that final organ size is a function of the amount of UGM originally present.

While the androgen dependency of prostatic development and growth has been known for some time (Jost, 1965; Price and Ortiz, 1965), androgen dependency now can be related to mesenchymal-epithelial interactions through analysis of tissue recombination experiments utilizing Tfm (testicular feminization) mice. Tfm mice are insensitive to androgens due to defective androgen receptors, and thus are feminized and completely lack prostates (Ohno, 1979). The four possible tissue recombinants possible between Tfm and wild-type tissues were constructed and exposed to physiological levels of androgens as a result of grafting the tissue recombinants into intact male hosts (Fig. 2). Prostatic morphogenesis occurred only in those recombinants prepared with wild-type mesenchyme (wild-type mesenchyme + wild-type epithelium or wild-type mesenchyme + Tfm epithelium). In contrast, prostatic differentiation failed to occur in recombinants constructed with Tfm mesenchyme (Cunha and Lung, 1978; Lasnitzki and Mizuno, 1980; Cunha *et al.*, 1980b), even when wild-type epithelium was used. In comparable experiments with the embryonic mouse mammary gland (Kratochwil and Schwartz, 1976; Drews and Drews, 1977), it has been shown that androgen-induced epithelial regression requires an androgen responsive (wild-type) mesenchyme. Thus, in both the developing prostate and embryonic mouse mammary gland, mesenchyme is the actual target and mediator of androgenic effects upon the epithelium. This concept is corroborated by the finding of nuclear ^{3}H-DHT binding sites in mesenchymal cells of wild-type UGS and mammary gland using steroid autoradiography (Shannon *et al.*, 1981; Shannon and Cunha, 1983; Takeda *et al.*, 1985; Wasner *et al.*, 1983). Tfm UGM lacks nuclear ^{3}H-DHT binding sites (Cunha *et al.*, 1982), and recent studies demonstrate that Tfm tissues lack mRNA encoding the androgen receptor (Lubahn *et al.*, 1988).

Prostatic development in tissue recombinants composed of wild-type UCM associated with either Tfm UGE or Tfm bladder epithelium (BLE) involves three major processes: ductal morphogenesis (the formation of a branched network of epithelial ducts), epithelial proliferation, and secretory cytodifferentiation. all of these processes are androgen-dependent and are expressed in Tfm epithelium, which remains devoid of androgen receptors even after prostatic induction (Cunha *et al.*, 1983c;

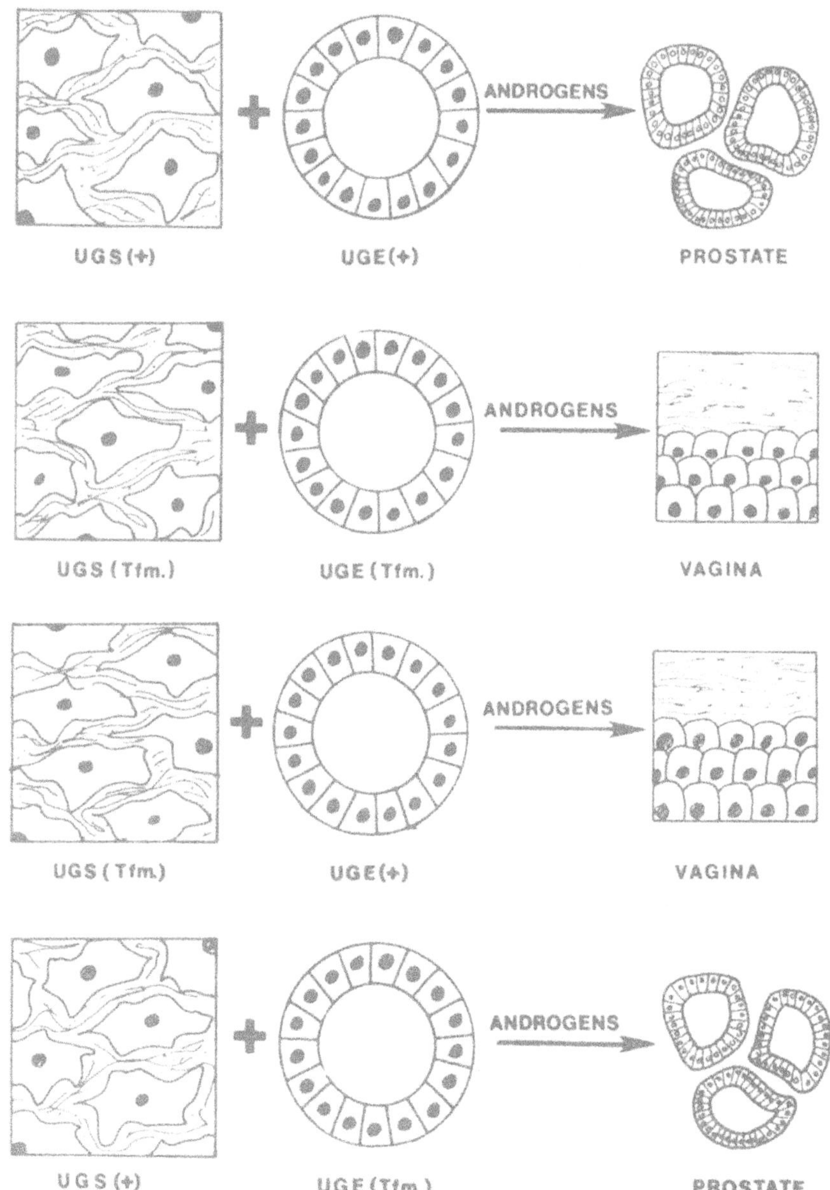

Figure 2. Summary of tissue recombinants studies between urogenital sinus epithelium and mesenchyme from Tfm (testicular feminization) and wild-type mice (from Cunha *et al.*, 1980b).

Shannon and Cunha, 1984). This demonstrates that androgen-receptor positive stromal cells play a central role in regulating these "androgen-induced" events within the epithelium (Cunha *et al.*, 1983c; Shannon and Cunha, 1984; Sugimura *et al.*, 1986b).

Since the initial analyses of the Tfm/wild-type model focused upon prostatic organogenesis, it could be argued that these conclusions might not be relevant to the mature prostate. However, when wild-type UGM + Tfm BLE recombinants are grown for 30 or more days, mature prostatic tissue forms which can be compared with wild-type prostate or tissue recombinants composed of wild-type UGM + wild-type BLE (Cunha *et al.*, 1980c; Neubauer *et al.*, 1983). Proteins produced by wild-type UGM + Tfm BLE recombinants are remarkably similar to those of wild-type prostate as judged by two dimensional gel electrophoresis (Cunha and Chung, 1981). Prostatic epithelial cells of wild-type UGM + Tfm BLE and wild-type UGM + wild-type BLE recombinants undergo similar regressive changes as judged histologically in response to androgen deprivation even though epithelial androgen receptors are present only in specimens prepared with wild-type (But not Tfm) epithelium (Cunha and Chung, 1981; Sugimura *et al.*, 1986b). Finally, androgen-induced DNA synthesis, as judged by both autoradiographic and biochemical procedures, is comparable in prostates that are either completely wild-type or are composed of wild-type mesenchyme +Tfm epithelium (Cunha and Chung, 1981; Sugimura *et al.*, 1986b). Thus, several lines of evidence emphasize the concept that epithelial growth and differentiation are regulated by androgens indirectly through androgen-dependent mediators of stromal origin.

RELEVANCE TO CANCER AND OTHER PROLIFERATIVE DISEASES

The rate of epithelial proliferation is determined by factors either intrinsic or extrinsic to the epithelium itself. This raises the possibility that proliferative disorders of the epithelium may have an etiology based upon an altered interaction with its stroma. In this regard, the histopathological observations of McNeal (1978) on the human prostate in combination with our studies on the role of mesenchyme as an inducer of prostatic growth and development have lead to the hypothesis that the formation of new ductal-acinar tissue in benign prostatic hypertrophy (BPH) is due to a re-activation of embryonic-like inductive activity in BPH stroma. In a similar fashion, epidermal hyperplasia in psoriasis has been shown to be due, in part, to the growth promoting activity of the dermis which suggests that fibroblasts associated with hyper-proliferative epithelium are themselves abnormal (Saiag *et al.*, 1985). In this regard, the fibroblasts associated with emerging carcinomas exhibit a variety of abnormalities and have been postulated to play a role in the carcinogenetic process (Chaudhuri *et al.*, 1975; Oishi *et al.*, 1981; Bauer *et al.*, 1977; Hodges *et al.*, 1978). Indeed, polyoma-induced carcinogenesis of the salivary gland occurs at a vastly higher rate under circumstances in which the epithelium is associated with mesenchyme as opposed to when the epithelium is grown by itself (Dawe *et al.*, 1968).

Once carcinoma cells emerge, they may retain a responsiveness to inductive influences from mesenchyme. De Cossa *et al.*, (1973, 1975) reported that mammary carcinoma cells exhibit a more orderly histo-differentiation and exhibit a lower growth rate when grown in association with embryonic mammary mesenchyme. More recently, Sakakura *et al.*, (1979c, 1981) have shown that certain embryonic mesenchymes accelerate the development of mammary carcinomas. Cooper and Pinkus (1977) demonstrated that when basal cell carcinomas are grow in association

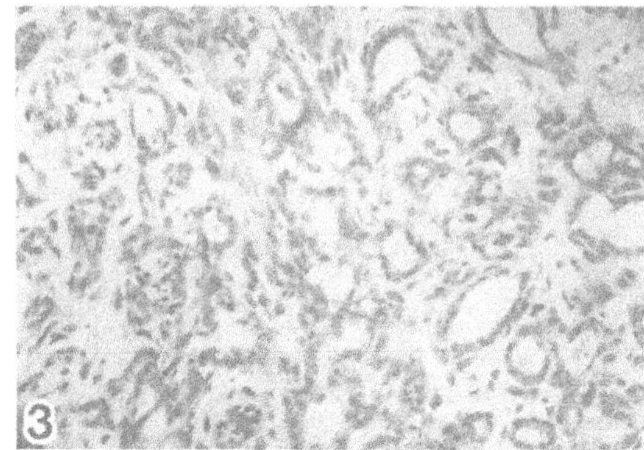

Figure 3. Histological section of the R3327 Dunning prostate adenocarcinoma. Note the small narrow tubules of undifferentiated epithelial cells (250X).

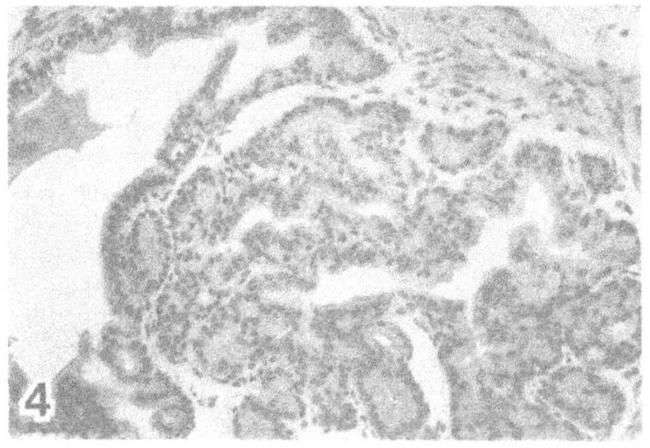

Figure 4. Histological section of Dunning tumor epithelial cell grown in association with urogenital sinus mesenchyme for 1 month under the renal capsule of a male host. Note that the epithelium is tall columnar and secretory (250X).

with normal stroma the malignant epithelial cells differentiated and lost their malignant properties. Similar methodology has been used to study epithelial-mesenchymal interactions in human oral mucosal lesions (MacKenzie *et al.*, 1979). Transitional carcinomas of the urinary bladder can be induced by urogenital sinus mesenchyme (UGM) to change into adenocarcinomatous structures (Fujii *et al.*, 1982). These observations provide ample evidence for a continued influence of connective tissue on carcinoma cells.

Recently, we have examined the influence of UGM on the R3327 Dunning rat prostatic adenocarcinoma. This tumor, which has been transplanted serially in male hosts since its discovery (Dunning, 1963), has maintained a stable histopathological phenotype, being composed of small ducts lined by a simple squamous to simple cuboidal epithelium (Isaacs, 1987; Smolev *et al.*, 1977a,b; Coffey *et al.*, 1979) (Fig. 3). Its growth rate is rather low having a doubling time of about 3 weeks (Weisman *et al.*, 1977). In our experiments 0.5-1mm^3 pieces of Dunning tumor (DT) were transplanted either alone or in combination with UGM and grown under the kidney capsule of syngeneic hosts for 1 month. An obvious effect of the mesenchyme was observed as the grafts of UGM + DT were substantially larger than the grafts of the UGM or the Dunning tumor by themselves. Histologically, the undifferentiated tumor cells in UGM + DT recombinants became highly differentiated tall columnar secretory epithelial cells (Fig. 4). This remarkable change in phenotype raises the possibility that the undifferentiated tumor cells of the parental tumor had terminally differentiated and thus might no longer be malignant. Studies in progress reassessing this issue and also the effect of UGM on the rate of proliferation of the Dunning tumor cells.

In summary, evidence is summarized which demonstrates that epithelial-mesenchymal interactions, which are fundamental during organogenesis, continue to play important roles in adulthood. A broad spectrum of adult epithelial cells exhibit a striking developmental plasticity and can be induced to express completely new morphological and functional properties by mesenchyme. In addition, certain adult stomas have been shown to have both instructive and permissive effects on epithelial morphogenesis and growth. These findings emphasize the concept that epithelial-stromal interactions remain operative and are crucially important in adulthood. Of necessity this means that proliferative disorders and specifically carcinogenesis may be related to abnormalities in the signaling between epithelium and stroma in adulthood. Future progress on the paracrine mediators that are involved in these cell-cell interactions will surely expand our knowledge of the causes of cancer and may lead to new therapeutic strategies.

ACKNOWLEDGEMENTS

The work presented in this report was supported by grants DK32157, HD21919 and CA05388 from the NIH.

REFERENCES

Bauer E.A., Gordon J.M., Reddick M.E. and Eisen A.Z. 1977. Quantitation and immunocytochemical localization of human skin collagenase in basal cell carcinoma. J. Invest. Derm. 69:363.

Bernimoulin J.P. and Schroeder H.E. 1980. Changes in the differentiation pattern of oral mucosal epithelium following heterotopic connective tissue transplantation in man. Path. Res. Pract. 166:290.

Billingham, R.E. and Silvers, W.K. 1968. Dermoepidermal interactions and epithelial specificity. In: Epithelial-Mesenchymal Interactions (R. Fleischmajer and Billingham R.E., eds.), Williams and Wilkins, Baltimore, p. 252.

Bissell M.J., Lee E.Y.H., Li M., Chen L. and Hall G. 1987. Role of extracellular matrix and hormones in modulation of tissue-specific functions in culture: mammary gland as a model for endocrine sensitive tissues. In: Benign Prostatic Hyperplasia: II (H. Rodgers, S. Coffey, G. Cunha, T. Grayhack, T. Hinman, Jr., and R. Horton, eds.), U.S. Government Printing Office, Washington, D.C., p. 39.

Blum J.L., Zeigler M.E. and Wicha M.S. 1987. Regulation of rat mammary gene expression by extracellular matrix components. Exp. Cell Res. 173:322.

Botero A., Ruben M.P. and Kramer G.M. 1975. connective tissue grafts: Induction of formation of gingiva in mucosal receptor sites. J. Periodont. 46:630.

Briggaman R.A. 1982. Epidermal-dermal interactions in adult skin. J. Invest. Dermatol. 79:21.

Chaudhuri S., Koprowska I. and Rowinski J. 1975. Different agglutinability of fibroblasts underlying various precursor lesions of human uterine cervical carcinoma. Cancer Research 35:2350.

Chung L.W.K. and Cunha G.R. 1983. Stromal-epithelial interactions. II. Regulation of prostatic growth by embryonic urogenital sinus mesenchyme. Prostate 4:503.

Chung L.W.K., Matsuura J. and Runner M.N. 1984. Tissue interactions and prostatic growth. Biol. Reprod. 31:155.

Coffey D.S., Isaacs J.T. and Weisman R.M. 1979. Animal models for the study of prostatic cancer. In: Prostatic Cancer (G.P. Murphy, ed.), PSG Publishing, Boston, p. 89.

Cohen S. 1965. The dermal papilla. In: Biology of the Skin and Hair Growth (A.G. Lyne and B.F. Short, eds.), Angus and Robertson, Sydney, p. 183.

Cohen J. 1969. Dermis, epidermis and dermal papillae interacting. In: Advances in Biology of Skin: Hair Growth, Vol 9 (W. Montagna and R.L. Dobson, eds.), Pergamon Press, New York, p. 1.

Cooke P.S., Fujii D.K. and Cunha G.R. 1987. Vaginal and uterine stroma maintain their inductive properties following primary culture. *In Vitro* Cellular and Develop. Biol. 23:159.

Cooper M. and Pinkus H. 1977. Intrauterine transplantation of rat basal cell carcinoma: A model for reconversion of malignant to benign growth. Cancer Research 37:2544.

Cunha G.R. 1972. Epithelio-mesenchymal interactions in primordial gland structures which become responsive to androgenic stimulation. Anat. Rec. 172: 179.

Cunha G.R. 1975. Age-dependent loss of sensitivity of female urogenital sinus to androgenic conditions as a function of the epithelial-stromal interaction in mice. Endocrinology 95:665.

to the studies involving the control of neoplastic growth should greatly assist the testing of MIS as a therapeutic tool for treatment of certain gynecologic cancers.

REFERENCES

Blanchard M. and Josso N. 1974. Source of anti-Mullerian hormone synthesized by the fetal testis: Mullerian inhibiting activity of fetal bovine Sertoli cells in tissue culture. Ped. Res. 8:968.

Brautigen D.L., Bornstein P. and Gallis B. 1981 Phosphotryrosyl-protein phosphate: specific inhibition by zinc ion. J. Biol. Chem. 256:6519.

Budzik G.P., Donahoe P.K. and Hutson J.M. 1985. A possible purification of Mullerian Inhibiting Substance and a model for its mechanism of action. In: "Developmental mechanism: normal and abnormal". (Lash, J.W. and Saxen, L. eds.), Alan R. Liss Inc., New York, p. 207.

Budzik G.P., Hutson J.M., Ikawa H. and Donahoe P.K. 1982. The role of zinc in Mullerian duct regression. Endocrinology. 110:1521.

Carpenter G., King L. and Cohen S. 1979. Rapid enhancement of protein phosphorylation in A-431 cell membrane preparations by epidermal growth factor. J. Biol. Chem. 254:4884.

Cate R.L., Mattaliano R.J., Hession C., Tizard R., Farber N.M., Cheung A., Ninfa E.G., Frey A.Z., Gash D.J., Chow E.P., Fisher R.A., Bertonis J.M., Torres G., Wallner B.P., Ramachandran K.L., Ragin R.C., Manganaro T.F., MacLaughlin D.T., and Donahoe P.K. 1986. Isolation of the bovine and human genes for Mullerian Inhibiting Substance and expression of the human gene in animal cells. Cell 45:685.

Cigarroa F., Coughlin J.P., Donahoe P.K., White M.F., Uitvlugt N. and MacLaughlin D.T. 1989. Recombinant Mullerian Inhibiting Substance inhibits epidermal growth factor receptor tyrosine kinase. Growth Factors. In press.

Cohen S., Chinkers M. and Ushero M. 1981. EGF-receptor protein kinase phosphorylates tyrosine and may be related to the transforming kinase of Rous sarcoma virus. Cold Spring Harbor Conf. Cell Prolif. 8:801.

Coughlin J.P., Donahoe P.K., Budzik G.P. and MacLaughlin D.T. 1987. Mullerian Inhibiting Substance blocks autophosphorylation of the EGF receptor by inhibiting tyrosine kinase. Mol. Cell. Endocrinol. 49:75.

Donahoe P.K., Cate R.L., MacLaughlin D.T., Epstein J., Fuller A.F., Takahashi M., Coughlin J.P., Ninfa E.G., and Taylor L.A. 1987. Mullerian Inhibiting Substance: gene structure and mechanism of action of a fetal regressor. Recent Progm. Horm. Res. 43:431.

Donahoe P.K., Fuller A.F., Scully R.E., Guy S.R., and Budzik G.P. 1981. Mullerian Inhibiting Substance inhibits growth of human ovarian cancer in nude mice. Ann. Surg. 194:472.

Donahoe P.K., Hutson J.M., Fallat M.E., Kamagata S. and Budzik G.P. 1984. Mechanism of action of Mullerian Inhibiting Substance. Ann. Rev. Physiol. 46:53.

Donahoe P.K., Swann D.A., Hayashi A. and Sullivan M.D. 1979. Mullerian

Cunha G.R. 1976. Stromal induction and specification of morphogenesis and cytodifferentiation of the epithelium of the Mullerian ducts and urogenital sinus during development of the uterus and vagina in mice. J. Exp. Zool., 196:361.

Cunha G.R. and Lung B. 1978. The possible influences of temporal factors in androgenic responsiveness of urogenital tissue recombinants from wild-type and androgen-insensitive (Tfm) mice. J. Exp. Zool. 205:342.

Cunha G.R., Lung B. and Reese B.A. 1980a. Glandular epithelial induction by embryonic mesenchyme in adult bladder epithelium of balb/c mice. Invest. Urol. 17:302.

Cunha G.R., Chung L.W.K., Shannon J.M. and Reese B.A. 1980b. Stomal-epithelial interactions in sex differentiation. Biol. Reprod. 22:19.

Cunha G.R., Reese B.A. and Sekkingstad M. 1980c. Induction of nuclear androgen-binding sites in epithelium of the embryonic urinary bladder by mesenchyme of the urogenital sinus of embryonic mice. Endocrinology 107:1767.

Cunha G.R. and Chung L.W.K. 1981. Stromal-epithelial interactions: Induction of prostatic phenotype in urothelium of testicular feminized (Tfm/y) mice. J. Steroid Biochem. 14:1317.

Cunha G.R., Shannon J.M., Taguchi O., Fujii H. and Chung L.W.K. 1982. Mesenchymal-epithelial interactions in hormone-induced development. J. of Animal Science 55 (Suppl.):14.

Cunha G.R., Fujii H., Neubauer B.L., Shannon M., Sawyer L.M. and Reese B.A. 1983a. Epithelial-mesenchymal interactions in prostatic development. I. Morphological observations of prostatic induction by urogenital sinus mesenchyme in epithelium of the adult rodent urinary bladder. J. Cell Biol. 96:1662.

Cunha G.R., Chung L.W.K., Shannon J.M., Taguchi O. and Fujii H. 1983b. Hormone-induced morphogenesis and growth: Role of mesenchymal-epithelial interactions. Recent Prog. Horm. Res. 39:559.

Cunha G.R. and Vanderslice K.D. 1984. Identification of species origin of mammalian cells in histological sections. Stain Technol. 59:7.

Cunha G.R., Higgins S.J. and Young P.F. 1988. Seminal vesicle mesenchyme induces heterotypic Wolffian duct derived epithelium to express seminal vesicle differentiation and to secrete seminal vesicle secretory proteins. Endocrinology 122 (Suppl.):189.

Daniel C.W. and DeOme K.B. 1965. Growth of mouse mammary glands *in vivo* after monolayer culture. Science 149:634.

Daniel C.W., DeOme K.B., Young J.T., Blair P.B. and Faulkin L.J. 1965. The *in vivo* life span of normal and neoplastic mouse mammary glands: A serial transplantation study. Proc. Natl. Acad. Sci. (USA) 61:52.

Daniel C.W. and Young L.J. 1971. Influence of cell division on an aging process. Life span of mouse mammary epithelium during serial propagation *in vivo*. Exp. Cell Res. 65:27.

Dawe C.J., Morgan W.D. and Slatick M.S. 1968. Salivary gland neoplasms in the role of normal mesenchyme during salivary gland morphogenesis.

In: Epithelial-Mesenchymal Interactions (R. Fleischmajer and R.E. Billingham, eds.), Williams and Wilkins Company, Baltimore, p. 295.

De Cosse J.J., Gossens C.L., Kuzma J.F. and Unsworth B.R. 1975. Embryonic inductive tissues that cause histological differentiation of murine mammary carcinoma *in vitro*. J. Natl. Cancer Inst. 54:913.

DeCosse J.J., Gossens C.L. and Kurma J.F. 1973. Breast Cancer: Induction of differentiation by embryonic tissue. Science 181:1057.

DeKlerk D.P. and Coffey D.S. 1978. Quantitative determination of prostatic epithelial and stromal hyperplasia by a new technique: Biomorphometrics. Invest. Urol. 16:240.

Donn B.H. 1978. The free connective tissue autograft: A clinical and histologic wound healing study in humans. J. Periodont. 49:253.

Drews U. and Drews U. 1977. Regression of mouse mammary gland anlagen in recombinants of Tfm and wild-type tissues: Testosterone acts via the mesenchyme. Cell 10:401.

Dunning W.F. 1963. Prostate cancer in the rat. Natl. Cancer Inst. Monogr. 12:351.

Ekblom P. 1984. Basement membrane proteins and growth factors in kidney differentiation. In: The Role of Extracellular Matrix in Development (R.L. Trelstad, ed.), Alan R. Liss, Inc., New York, p. 173.

Evan H.M. and Long A. 1920. The oestrous cycle in the rat. Anat. Rec. 18:241.

Fujii H., Cunha G.R. and Norman J.T. 1982. The induction of adenocarcinomatous differentiation in neoplastic bladder epithelium by an embryonic prostatic inductor. J. Urol. 128:858.

Fusenig N.E. 1986. Mammalian epidermal cells in culture. In: Biology of the Integument, Vol. 2 Vertebrates (J. Bereiter-Hahn, A.G. Matolsy and K.S. Richards, eds.), Springer-Verlag, Berlin-Heidelberg, p. 409.

Fusenig N.E., Breitkreutz D., Lueder M., Boukamp P. and Worst P.K.M. 1981. Keratinazation and structural organization in epidermal cell cultures. In: International Cell Biology (H.G. Schweiger, ed.), Springer-Verlag, Berlin-Heidelberg, p. 1004.

Haffen K., Kedinger M., Simon P.M. and Raul F. 1987. Mesenchyme-dependent differentiation of epithelial progenitor cells in the gut. J. of Ped. Gastroenterol. Nutr. 6:14.

Hicks R.M. 1975. The mammalian urinary bladder: An accommodating organ. Biol. Rev. 50:215.

Higgins J., Young P., Brody R. and Cunha G.R. 1989. Induction of functional cytodifferentiation in the epithelium of tissue recombinants. I. Homotypic seminal vesicle recombinants. Development (in press).

Hodges G.M., Hicks R.M. and Spacey G.D. 1977. epithelial-stromal interactions in normal and chemical carcinogen-treated adult bladder. Cancer Research 37:3720.

Ibrahim L. and Wright E.A. 1977. Inductive capacity of irradiated dermal papillae. Nature 265:733.

Isaacs J.T. 1987. Development and characteristics of the available animal model systems for the study of prostatic cancer. In: Current Concepts and Approaches to the study of Prostate Cancer (D.S. Coffey, W.A. Gardner, Jr., N. Bruchovsky, M.I. Resnick, and J.P. Karr, eds.), Alan R. Liss, Inc., New York, p. 513.

Jahoda C.A.B., Horne K.A. and Oliver R.F. 1984. Induction of hair growth by implantation of cultured papilla cells. Nature 311:560.

Jost A. 1965. Gonadal hormones in the sex differentiation of the mammalian fetus. In: Organogenesis (R.L. DeHaan and H. Urpsrung, eds.), Holt, Rinehart and Winston, New York, p. 611.

Karring T., Land N.P. and Loe H. 1975. The role of gingival connective tissue in determining epithelial differentiation. J. Periodontal Res. 10:1.

Kedinger M., Haffen K. and Simon-Assmann P. 1986. Control mechanisms in the ontogenesis of villus cells. In: Molecular and Cellular Basis of digestion (P. Desnuelle, H. Sjostrom and O. Noren, eds.), Elsevier, New York, p. 323.

Kollar E.J. 1972. The development of the integument: Spatial, temporal and phylogenetic factors. Am. Zool. 12:125.

Kollar E.J. 1986. Tissue interactions in development of teeth and related ectodermal derivatives. In: Developmental Biology A Comprehensive Synthesis (R.B.L. Gwatkin, ed.), Plenum Press, New York, p. 297.

Kratochwil K. and Schwartz P. 1976. Tissue interaction in androgen response of the embryonic mammary rudiment of mouse: Identification of target tissue of testosterone. Proc. Natl. Acad. Sci. USA 73:4041.

Kratochwil K. 1987. Tissue combination and organ culture studies in the development of the embryonic mammary gland. In: Developmental Biology: A Comprehensive Synthesis. (R.B.L. Gwatkin, ed.), Plenum Press, New York, p. 315.

Lasnitzki I. and Mizuno T. 1980. prostatic induction and interaction of epithelium and mesenchyme from normal wild-type and androgen-insensitive mice with testicular feminization. J. Endocrinol. 85:423.

Leblond C.P. and Cheng H. 1976. Identification of stem cells in the small intestine of the mouse. In: Stem cells of Renewing Cell Populations (A.B. Cairnie, P.K. Lala and D.G. Osmond, eds.), Academic Press, New York, p. 7.

Lee E.Y.H., Parry G. and Bissell M.J. 1984. Modulation of secreted proteins of mouse mammary epithelial cells by the collagenous substrate. J. Cell Biol. 98:146.

Lillie F.R. and Wang H. 1943. Physiology of development of the feather. VI. The production and analysis of feather chimaerae in fowl. Physiol. Zool. 16: 1.

Lillie J.H., MacCallum D.K. and Jepsen A. 1983. The role of defined extracellular matrices on the growth and differentiation of mammalian stratified squamous epithelium. In: Epithelial-Mesenchymal Interaction in Development (R.H. Sawyer and J.F. Fallon, eds.), Praeger., New York, p. 93.

Lubahn D.B., Joseph D.R., Sullivan P.M., Willard H.F., French F.S., and Wilson E.M. 1988. Cloning of human androgen receptor complementary DNA and localization to the X chromosome. Science 240:327.

Mackenzie J.C., Dabelsteen E. and Roed-Peterson B. 1979. A method for studying epithelial-mesenchymal interactions in human oral mucosal lesions. Scand. J. Dent. Res. 87:234.

McNeal J.E. 1978. Evolution of benign prostatic enlargement. Invest. Urology 15:340.

Michalopoulos G. and Pitot H.C. 1975. Primary culture of parenchymal liver cells on collagen membranes. Exp. Cell Res. 94:70.

Neubauer B.L., Chung L.W.K., McCormick K., Taguchi __, Thompson T.C. and Cunha G.R. 1983. Epithelial-mesenchymal interactions in prostatic development. II. Biochemical observations of prostatic induction by urogenital sinus mesenchyme in epithelium of the adult rodent urinary bladder. J. Cell Biol. 96:1671.

Neubauer B.L., Best K.L., Hoover D.M., Slisz M.L., Van Frank R.M. and Goode R.L. 1986. Mesenchymal-epithelial interactions as factors influencing male accessory sex organ growth. Fed. Proc. 49:2618.

Nieuwkoop P.D. 1973. The 'organisation centre' of the amphibian embryo, its origin, spatial organization and morphogenetic action. Adv. Morphogen. 10:1.

Norman J.T., Cunha G.R. and Sugimura Y. 1986. The induction of new ductal growth in adult prostatic epithelium in response to an embryonic prostatic inductor. Prostate 8:209.

Ohno S. 1979. Major Sex Determining Genes. Springer Verlag, New York.

Oishi K., Romijn J.C. and Schroeder F.H. 1981. The surface character of separated prostatic cells and cultured fibroblasts of prostatic tissue as determined by concanavalin-A hemadsorption. Prostate 2:11.

Oliver R.F. 1968. The regeneration of vibrissae--a model for the study of dermal-epidermal interactions. In: Epithelial-Mesenchymal Interactions. (R. Fleischmajer and R.E. Billingham, eds.), Williams and Wilkins Company, Baltimore, p. 267.

Potten C.S. 1983. Stem cells in epidermis from the back of the mouse. In: Stem Cells: Their Identification and Characterization (C.S. Potten, ed.), Churchill Livingstone, New York, p. 155.

Potten C.S. and Hendry J.H. 1983. Stem cells in murine small intestine. In: Stem Cells: Their Identification and Characterization (C.S. Potten, ed.), Churchill Livingstone, New York, p. 155.

Price D. and Ortiz E. 1965. The role of fetal androgens in sex differentiation in mammals. In: Organogenesis (R.L. DeHaan and H. Ursprung, eds.), Holt, Rinehart and Winston, New York, p. 629.

Reid L.M. and Jefferson D.M. 1984. Cell culture studies using extracts of extracellular matrix to study growth and differentiation in mammalian cells. In: Mammalian Cell Culture (J.P. Mather, ed.), Plenum Press, New York, p. 239.

Saiag P., Coulomb B., Lebreton C., Bell E. and Dubertret L. 1985. Psoriatic fibroblasts induce hyperproliferation of normal keratinocytes in a skin equivalent model *in vitro*. Science 230:669.

Sakakura T., Sakagami Y. and Nishizuka Y. 1979a. Persistence of responsiveness of adult mouse mammary gland to induction by embryonic mesenchyme. Develop. Biol. 72:201.

Sakakura T., Nishizuka T. and Dawe C.J. 1979b. Capacity of mammary fat pads of adult C3h/HeMs mice to interact morphogenetically with fetal mammary epithelium. J. Natl. Cancer Inst. 63:733.

Sakakura T., Sakagami Y. and Nishizuka Y. 1979c. Acceleration of mammary cancer development by grafting of fetal mammary mesenchymes in C3H mice. Gann. 70:459.

Sakakura T., Sakagami Y. and Nishizuka Y. 1981. Accelerated mammary cancer development by fetal salivary mesenchyme isografted to adult mouse mammary epithelium. J. Nat. Cancer Institute 66:953.

Saunders J.W. and Gasseling M.T. 1968. Ectodermal-mesenchymal interactions in the origin of limb symmetry. In: Epithelial-Mesenchymal Interactions (R. Fleischmajer and R.E. Billingham, eds.), Williams and Wilkins, Baltimore, p. 78.

Sawyer R.H. 1983. The role of epithelial-mesenchymal interactions in regulating gene expression during avian scale morphogenesis. In: Epithelial-Mesenchymal Interactions in Development (R.J. Sawyer and J.F. Fallon, eds.), Praeger, New York, p. 115.

Saxen, L., Ekblom, P. and Thesleff, I. 1980. Mechanisms of morphogenetic cell interactions. In: Development in Mammals, Vol. 4 (M.H. Johnson, ed.), Elsevier, New York, p. 161.

Shannon J.M., Cunha G.R. and Vanderslice K.D. 1981. Autoradiographic localization of androgen receptors in the developing urogenital tract and mammary gland. Anat. Rec. 199:232.

Shannon J.M. and Cunha G.R. 1983. Autoradiographic localization of androgen binding in the developing mouse prostate. Prostate 4:367.

Shannon J.M. and Cunha G.R. 1984. Characterization of androgen binding and deoxyribonucleic acid synthesis in prostate-like structures induced in testicular feminized (Tfm/Y) mice. Biol. Reprod. 31:175.

Smolev J., Heston W.D.W., Scott W.W. and Coffey D.S. 1977a. Characterization of the Dunning R-3327-H prostatic adenocarcinoma: An appropriate animal model for prostatic cancer. Cancer Trt. Reps. 61:273.

Smolev J.K., Coffey D.S. and Scott W.W. 1977b. Experimental models for the study of prostatic adenocarcinoma. J. Urol. 118:216.

Spearman R.I.C. 1974. Alteration of keratinization in mouse ear epidermis in recombinant grafts with tail dermis. Acta Anat. 89:195.

Sugimura Y., Cunha G.R. and Donjacour A.A. 1986a. Morphogenesis of ductal networks in the mouse prostate. Biol. Reprod. 34:961.

Sugimura Y., Cunha G.R. and Bigsby R.M. 1986b. Androgenic induction of deoxyribonucleic acid synthesis in prostate-like glands induced in the urothelium of testicular feminized (Tfm/y) mice. Prostate 9:217.

Takeda H., Miguno T. and Lasnitzki I. 1985. Autoradiographic studies of androgen-binding sites in the rat urogenital sinus and postnatal prostate. J. Endocrinol. 104:87.

Thompson T.C. and Chung L.W.K. 1986. Regulation of overgrowth and expression of prostatic binding protein in rat chimeric prostate gland. Endocrinol. 118: 2437.

Wang H. 1943. The morphogenetic functions of the epidermal and dermal components of the papilla in feather regeneration. Physiol. Zool. 16: 325-350.

Wasner G., Hennermann I. and Kratochwil K. 1983. Ontogeny of mesenchymal androgen receptors in the embryonic mouse mammary gland. Endocrinology 113: 1771.

Weisman R.M., Coffey D.S. and Scott W.W. 1977. Cell kinetics studies of prostatic cancer: adjuvant therapy in animal models. Oncology 34:133.

MULLERIAN INHIBITING SUBSTANCE: STUDIES ON ITS MECHANISM OF ACTION AND ACTIVITY AS AN ANTI-TUMOR AGENT

David T. MacLaughlin*,
James Epstein** and
Patricia K. Donahoe**

*Vincent Research Laboratory
and
**Pediatric Surgical Research Laboratory
Massachusetts General Hospital
Boston, MA 02114

The urogenital ridge of the developing mammalian embryo is a complex structure in which the Wolffian and Mullerian ducts, each comprised of luminal epithelium surrounded by mesenchyme, develop side by side. As the maturing gonad begins to declare its genetic sex as either an ovary or testis, significant changes occur in the ducts contained within the urogenital ridge. If the gonad is an ovary the Wolffian duct, which develops into the epididymis, vas deferens, and seminal vesicles of the male, begins to atrophy while the Mullerian ducts expand. Conversely, testicular development coincides with the regression of the Mullerian ducts, the precursor of the ovarian coelomic epithelium, Fallopian tubes, uterus, cervix, and upper third of the vagina. As expected in this case, the Wolffian duct continues to enlarge.

It was the pioneering work of Jost and co-workers using a gonadectomized rabbit embryo model, nearly a half a century ago, which demonstrated that the process of Wolffian duct regression in females was passive, that is, dependent only upon the lack of the testicular hormone testosterone. In males, however, the process leading to the disappearance of the Mullerian duct was active and not simply the result of insufficient estrogen from an ovary. Furthermore, testosterone, when administered to the orchidectomized animals, could not obliterate the Mullerian ducts in the embryo. Jost proposed that testicular products other than testosterone, an anti- Mullerian hormone, must be responsible for the active regression of the Mullerian structures. Since the Wolffian component of the urogenital ridge withers in the absence of androgen, no such ovarian hormone was considered necessary. Jost's hypothesis proved to be correct when two separate laboratories isolated and purified such a glycoprotein hormone from neonatal bovine testicular extracts. This protein, termed Mullerian Inhibiting Substance (MIS) by our laboratory (for review see Donahoe *et al.*, 1987) and anti-Mullerian hormone by Josso and coworkers (1977), is a disulfide bond-linked 140 kDa homodimer that causes regression of fetal rat Mullerian ducts in an organ culture system which was developed by Picon *et al.*, (1969).

Upon disulfide bond reduction, the predominant species of MIS is a 70-74 kDa monomer. Also present under these conditions is a considerably smaller amount of a 50-56 kDa fragment. The nature of this minor species of MIS remains to be firmly established but preliminary evidence, discussed below, suggests that MIS may undergo site specific proteolytic cleavage. Amino-acid and carbohydrate compositional analyses indicated that MIS is rich in hydrophobic amino acids and contains approximately 13% carbohydrate by weight (Cate *et al.*, 1986; Picard *et al.*, 1986). Immunohistochemical studies showed that MIS is produced by fetal Sertoli cells (Hayashi *et al.*, 1984). In addition to causing Mullerian duct regression in the male embryo, MIS has recently been implicated as being involved in testicular descent (Hutson and Donahoe, 1986). More recent experiments have identified MIS in adult ovarian granulosa cells (Vigier *et al.*, 1984; Necklaws *et al.*, 1986; Takahashi *et al.*, 1986a) where it acts as an inhibitor of oocyte meiosis (Takahashi *et al.*, 1986b; Ueno *et al.*, 1988).

In collaboration with the Biogen Corporation (Cambridge, MA.), bovine cDNA and human genomic sequences for MIS were characterized and cloned by stable transfection into Chinese hamster ovarian (CHO) cells (Cate *et al.*, 1986). The mature human MIS protein is encoded for by a gene containing five exons and four introns. The addition of carbohydrate at two consensus glycosylation sites raises the molecular weight of the 56 kDa nascent protein to the 70 kDa monomer observed with the naturally occurring molecule under reducing conditions. Disulfide bonds link the monomers together, resulting in the formation of the 140 kDa dimeric form of MIS originally seen with the purified bovine testicular hormone. The human recombinant MIS in addition to possessing the chemical characteristics of the testicular protein, also causes Mullerian duct regression in the standard 14-1/2 day fetal rat urogenital ridge bioassay. An examination of the deduced primary amino acid sequences of bovine and human MIS revealed an overall high degree of homology (78%) with 108 of the final 112 residues identical. The bovine protein was 16 residues longer than the 535 amino acid human product. The nucleotide and amino acid sequences for bovine MIS determined with a different experimental approach agrees well with those of our group (Picard *et al.*, 1986). Comparison of the primary sequence of MIS with other known proteins reveals a structural homology with a number of biological response modifiers including transforming growth factor beta, inhibin and activin (Cate *et al.*, 1986), the decapentaplegia complex of *Drosophila melanogaster* and the *Xenopus laevis* Vgl protein. The nearly 30% homology observed is located in the C-terminal end of MIS and these other proteins, particularly adjacent to the cysteine residues involved in inter-chain disulfide bond formation. These proteins are now considered to be members of a super-gene family, perhaps evolving from a common ancestral precursor.

Clues to the mechanism of action of MIS come from the observation that epidermal growth factor (EGF), a hormone known to act via induction of autophosphorylation of tyrosine residues in its receptor, inhibits the action of MIS on the urogenital ridge (Hutson *et al.*, 1984). MIS, on the other hand, blocks the proliferative effect of EGF on a human cancer cell line, A-431 (Coughlin *et al.*, 1987).

In this report we summarize the recent work of our laboratory to elucidate the molecular mechanism of action of MIS and to determine if this fetal regressor may play a useful role as a chemotherapeutic agent for tumors of tissues of Mullerian origin. All of the experimental protocols and reagents used in the experiments described herein have been outlined previously in detail. Accordingly, the descriptions of particular studies are abbreviated to include features that are deemed of critical importance.

ONCOLOGY STUDIES

Experiments were designed to test whether MIS, a fetal regressor of Mullerian ducts, may also block the proliferation of malignant tumors arising from structures of Mullerian origin. They were performed with three different protocols.

In a microcytotoxicity assay, established cell lines including HOC-1, a human ovarian carcinoma, were grown on microtiter plates in the presence or absence of preparations of bovine or human recombinant MIS (Donahoe *et al.*, 1979). Cell number and viability were assessed in the treatment groups and compared to the control incubations in which the cells were exposed to buffer or testicular proteins devoid of MIS activity in the standard organ culture assay. For comparison, the effect of MIS on the proliferation of human foreskin fibroblasts and a glioblastoma cell line was also studied.

Clonogenic stem cell assays were also performed in which the ability of MIS to block the formation of colonies of cells in soft agar was measured according to a stem cell assay described earlier (Fuller *et al.*, 1982, 1985). In these experiments established human cancer cell lines as well as cells harvested from 43 fresh surgical samples were suspended in soft agar and incubated for up to 21 days with MIS, appropriate buffer controls and protein negative controls or chemotherapeutic agents as positive controls. At the conclusion of the experiments, colonies of cells were counted and the effect of MIS on colony formation was determined relative to the positive and negative control cultures.

The final protocol employed to assess the anti-tumor actions of MIS was to determine the effect of this hormone on the growth of a human ovarian carcinoma HOC-21 (Donahoe *et al.*, 1981) or an endometrial carcinoma SCRC-1 (Fuller *et al.*, 1984) implanted into nude mice. The cancer cell lines were injected sc into Balb/c homozygous nude mice after pre-incubation with partially purified bovine MIS. Animals were inspected daily for tumor growth, and the earliest day of appearance was noted. At selected times throughout the study, animals were sacrificed and tumor size and histology were evaluated. In each study group, which included buffer and equiprotein negative controls, MIS treatment and chemotherapeutic agent arms, the proportion of animals free of tumor was determined and plotted versus time since inoculation. A human colonic cancer cell line (SW-48) was also tested as a non-Mullerian neoplasia.

EGF RECEPTOR AUTOPHOSPHORYLATION ASSAY

The effect of MIS on EGF-induced autophosphorylation of the EGF receptor was assessed in two separate systems both employing a human vulvar carcinoma cell line A-431. This cell line, rich in EGF receptor, has been studied by this laboratory previously (Coughlin *et al.*, 1987, Ciggaroa *et al.*, 1989). In one set of studies, bovine and human recombinant MIS were added in the presence or absence of EGF directly to plasma membrane fraction of A-431 cells in the S phase of the cell cycle prepared according to the protocol of Carpenter *et al.*, (1979, 1981). After a 10 minute incubation with ^{32}P ATP, the autophosphorylation of the EGF receptor was determined by polyacrylamide gel electrophoresis and densitometry of the autoradiographs of the 170 kDa EGF receptor complex.

Alternatively, similar experiments were done on intact A-431 cells grown in microlayer culture in phosphate-free media. At the conclusion of the incubation with hormone and ^{32}P orthophosphate cells were lysed, the membranes solubilized and the radiolabeled EGF receptor precipitated with anti-EGF receptor or anti-phosphotyrosine antibody prior to electrophoresis and autoradiography (Cigarroa *et al.*, 1989).

In all experiments the specificity of the MIS effect was documented by performing the protocols with equiprotein fractions from the MIS ion exchange-dye affinity or immunoaffinity purification schema that did not contain MIS activity in the standard regression bioassay. Furthermore, all buffers were tested in these phosphorylation assays to rule out a non-specific effect. Finally, the activity of MIS as an inhibitor of EGF-induced receptor autophosphorylation was tested in the presence of an MIS-specific monoclonal antibody as yet another way of evaluating the effect as MIS dependent. Other details are listed in the text or legends to the figures.

THE ANTI-PROLIFERATIVE EFFECTS OF MIS

In order to test the attractive hypothesis that a naturally occurring fetal regressor of Mullerian ducts may possess similar activity against tumors arising from tissues of Mullerian origin, several approaches were undertaken. The first was to determine if MIS could alter the proliferation pattern of established human cancer cell lines *in vivo* and *in vitro*. Accordingly, human ovarian HOC-21 (Yamada, 1974) and endometrial (SCRC-1) cancers were grown in the presence or absence of biologically active MIS in microcytotoxicity, clonogenic and nude-mouse implant studies (Donahoe *et al.*, 1979, 1981; Fuller *et al.*, 1982, 1984, 1985). The results of these studies (Table I) demonstrate clearly that the partially purified bovine MIS inhibited the growth of these cell lines regardless of the assay employed. Purification fractions free of MIS, as measured by bioassay, were inactive in this regard, and normal or cancer cells from non-Mullerian sources were unaffected by incubation with MIS.

Table I. Anti-proliferative Effect of MIS on Human Cell Lines

	MONOLAYER		CLONOGENIC		NUDE MOUSE	
Cells	Mis	Control	MIS	control	MIS	control
HOC21	+	-	+	-	+	-
SCRC1	nd	nd	nd	nd	+	-
Fibroblasts	-	-	nd	nd	nd	nd
Glioblastoma	-	-	nd	nd	nd	nd
Colon Cancer SW48	nd	nd	-	-	-	-

+: indicates MIS inhibited cell growth
control: indicates equiprotein fraction without measurable MIS
nd: not determined

In another series of studies, human tumor cells were prepared from 43 surgical specimens from 40 patients and plated into a standard clonogenic stem cell assay (Fuller *et al.*, 1985). Twenty six (60%) of the tumors grew sufficiently to be tested with MIS, and 23/26 (88%) were significantly inhibited when compared to buffer and non-MIS protein treated controls (Figure 1).

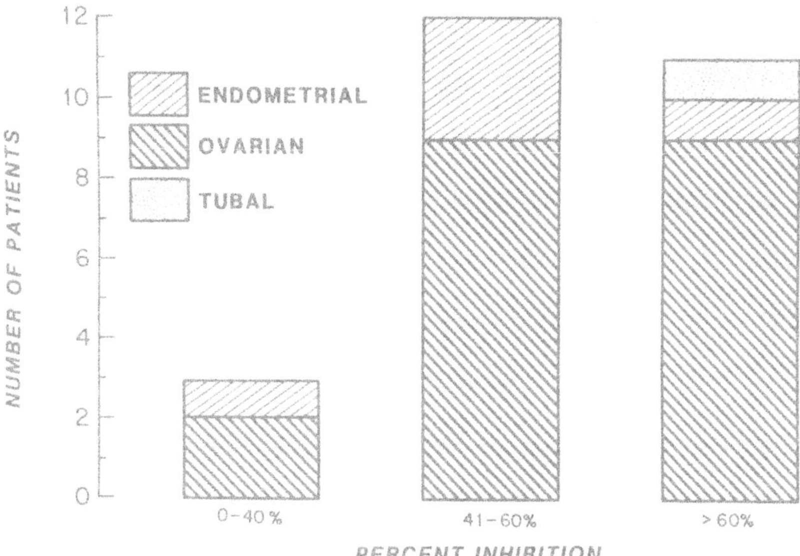

Figure 1. Colony inhibition of tumors from patients with gynecologic cancer. Inhibition by bovine MIS of colony formation of human cancer cells obtained from surgical specimens of a variety of gynecologic tumors. Twenty three of 26 cases showed significant inhibition when incubated with MIS as compared to the phosphate buffered saline controls. Eleven of the 26 cases showed greater than 60% inhibition of colony growth (Fuller *et al.*, 1985 with permission).

Taken together, these data support the hypothesis that MIS may indeed inhibit the proliferation of human malignancies though several critical studies remain to be completed. All of the results summarized thus far were obtained using highly purified, but not homogeneous bovine MIS. Unpublished studies using these and other assays show pure human recombinant MIS to be less active. Namely, it is ineffective against established cell lines, but remains active when treating tumor explants. Among the explanations for the decrease in activity as the degree of purity increases is the possibility that MIS, as produced by the cloning technique, exists as an inactive pro-hormone which must be processed to possess activity and that by immunoaffinity purification the inactive form predominates. Recent evidence (Pepinsky *et al.*, 1988; Cate *et al.*, in preparation) demonstrates that the intact 140 kDa dimer of MIS must be proteolytically cleaved to demonstrate its biological action. The specific protease plasmin cleaves human MIS at a monobasic cleavage site at amino acid residue 427 to yield a 30 kDa disulfide-linked dimer of the C-terminal region of MIS and a 110 kDa intact dimer of the N-terminal portion of MIS. Purified N- and C-terminal MIS fragments prepared from plasmin-treated, normal recombinant MIS are inactive when added separately to the bioassay; however, co-addition of these fragments restores the ability of MIS to cause regression of the Mullerian duct in the fetal rat urogenital ridge. It will be of interest, therefore, to re-evaluate the *in vitro* anti-proliferative effects of the pure recombinant human MIS with C- and N-terminal fragments of MIS. These preliminary experiments indicate that MIS must be proteolytically cleaved but with the N- and C-fragment non-covalently associated to be fully active. Studies are now underway to determine if experimentally cleaved MIS or its isolated fragments can reproduce the experimental results obtained with the naturally occurring protein.

MIS INHIBITS AUTOPHOSPHORYLATION OF THE EGF RECEPTOR

Our current model of the molecular mechanism of action of MIS has evolved from a number of control studies in which the effect of various constituents of buffers used to purify MIS from neonatal bovine testis extracts were assessed in the standard organ culture bioassay system (Picon *et al.*, 1969). The divalent metal chelator EDTA mimicked MIS action as a regressor of Mullerian ducts in the fetal rat urogenital ridge. Only zinc was found to counteract this effect. In addition, coincubation of zinc with MIS completely inhibited MIS in a dose dependent manner (Budzik *et al.*, 1982).

The report of Brautigen *et al.* (1981), describing a membrane bound phosphotyrosyl protein phosphatase inhibited by zinc resulted in an examination of the effect of MIS on hormones known to induce protein tyrosyl phosphorylation. Epidermal growth factor, EGF, a factor which causes the autophosphorylation of its receptor at specific tyrosine residues (Cohen *et al.*, 1981) also completely inhibited Mullerian duct regression by MIS (Donahoe *et al.*, 1984). Other growth factors including nerve growth factor, insulin, platelet-derived growth factor (PDGF) (Budzik *et al.*, 1985) and, more recently, transforming growth factor β (TGF-B) were without effect on MIS activity or duct regression.

The MIS-EGF interaction was then further pursued as a model system with which to study the role of membrane phosphorylation state n the action of MIS. Since EGF is an inhibitor of MIS, experiments were undertaken to determine if MIS could block the action of EGF. The EGF-induced proliferation of the EGF receptor-rich human cancer cell line A-431 was blocked by the addition of MIS to a standard clonogenic assay (Coughlin *et al.*, 1987). Subsequent experiments focused on the influence of MIS on EGF action at the molecular level, namely the induction of EGF receptor tyrosine kinase activity and the resulting autophosphorylation of the EGF receptor. Studies using both bovine and human recombinant MIS performed on monolayer cell cultures of A-431 cells and isolated A-431 cell plasma membrane demonstrated a clear inhibition of EGF stimulated receptor autophosphorylation. Representative results are given in Figures 2 and 3. this effect was not due to MIS hydrolyzing the substrate, nor was it due to an alteration of the number of EGF receptor binding sites (Coughlin *et al.*, 1987). The MIS effect was not reversible by a large molar excess of ATP, EGF or its required cofactor manganese (Cigarroa *et al.*, 1989).

The findings just summarized have all been obtained using partially purified bovine and human recombinant MIS. Control media, taken from the CHO cell line in which the human MIS gene was transfected, fails to block EGF activity. The effect of MIS-containing preparations is inhibited by co-incubation of MIS with a specific anti-MIS monoclonal antibody (Figure 4). Human recombinant MIS purified essentially to homogeneity by immuno-affinity chromatography, however, fails to demonstrate the anti-EGF action of MIS. These latter preparations of MIS retain the ability to induce Mullerian duct regression in the standard rat urogenital ridge bioassay.

In spite of these results, however, all MIS preparations that block EGF receptor autophosphorylation also cause Mullerian duct regression and EGF receptor autophosphorylation inhibition is only observed with preparations containing measurable MIS. At present, several possible explanations for decreased activity of purified recombinant MIS include the denaturation of the protein following chaotropic salt elution from the affinity columns used for purification and the removal of cofactors necessary for full expression of biological activity of the hormone.

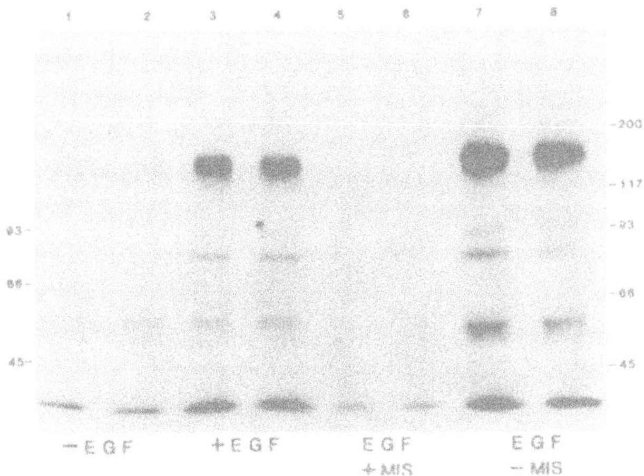

Figure 2. Bovine MIS inhibits the phosphorylation of 170 kDa EGF receptor in A-431 cell membrane. This autoradiograph of a 7.5% polyacrylamide gel shows the eGF dependent increase in receptor phosphorylation (lanes 1,2 compared to 3,4). Addition of 200 nM bovine MIS completely blocked the EGF effect (lanes 5,6). Equiprotein protein purification fractions devoid of MIS as measured by bio- and immunoassay were without effect (lanes 7,8). Molecular weight markers re given on the right and left sides of the figure (Coughlin *et al.*, 1987 with permission).

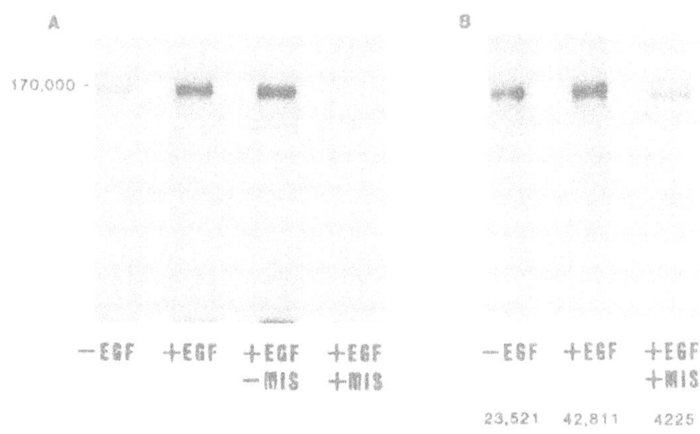

Figure 3. Human recombinant MIS inhibits EGF receptor autophosphorylation. (A) Human recombinant MIS (70 nM) purified from serum-containing culture media of CHO cells transfected with the MIS gene inhibits EGF induced EGF receptor autophosphorylation in A-431 cell membranes (compare +EGF vs. +EGF+MIS). The addition of non-MIS protein purified from the same media had no effect on the EGF response (+EGF-MIS). (B) Human recombinant MIS (100 nM) purified from serum free conditioned CHO media and therefore not co-purified with non-CHO cell protein, similarly blocks EGF receptor autophosphorylation (+EGF+MIS). The values given at the bottom of each lane in panel B are the quantitative densitometry data for this autoradiograph documenting the complete inhibition of the EGF induced elevation in receptor phosphorylation (Cigarroa *et al.*, 1989 with permission).

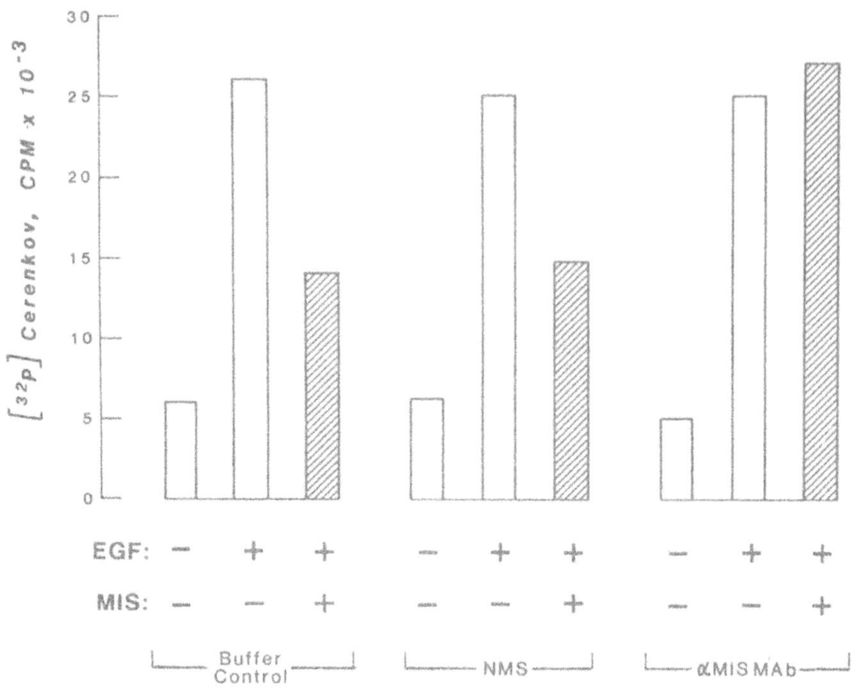

Figure 4. MIS Immunoabsorption. Human recombinant MIS purified from serum-containing conditioned media was preincubated with antibody buffer control, normal mouse serum (NMS), or monoclonal anti-human MIS antibody (MISMAb), and then tested in the A-431 cell membrane phosphorylation assay. MIS (+EGF+MIS) pre-incubated with buffer control and normal mouse serum inhibited EGF-stimulated autophosphorylation of its receptor (+EGF-MIS). In contrast, MIS (+EGF+MIS) pre-incubated with monoclonal anti-human MIS antibody lost its ability to inhibit EGF-stimulated autophosphorylation of its receptor. Basal EGF receptor phosphorylation is shown (-EGF-MIS). Cerenkov counts were obtained for the 170 kDa EGF receptor isolated by polyacrylamide gel electrophoresis (Cigarroa *et al.*, 1989 with permission).

Since this preparation causes Mullerian duct regression, the MIS may renature in the three day organ culture assay or be resupplied with cofactors in the incubation medium or in cellular secretions. A more intriguing possibility has already been mentioned in relation to the anti-proliferative assays of MIS. Whether a fragment of MIS, that is the N- or C-terminal portion, or a combination of these species in proteolytically cleaved MIS is responsible for the inhibition of EGF induced tyrosine kinase activity is now the subject of intense scrutiny in this laboratory.

The present interpretation of MIS action, therefore, assumes interaction of MIS with its receptor followed by the inhibition of autophosphorylation of EGF receptor. The ensuing proliferative effects of EGF distal to the hormone-receptor complex are then blocked. Other factors, such as the gonadal estrogens which inhibit MIS activity (Hutson *et al.*, 1985) and the progestins and androgens which augment it (Ikawa *et al.*, 1982), as well as the post-translational modification of the MIS molecule discussed above, must be considered as key regulators of this process. Alteration of the intricate interplay of these hormones could cause aberrant reproductive tract maturation. The application of the knowledge of how MIS normally functions under physiologic circumstances

duct regression in the embryo correlated with cytotoxic activity against human ovarian cancer. Science 205:913.

Fuller A.F., Budzik G.P., Krane I.M. and Donahoe P.K. 1984. Mullerian Inhibiting Substance inhibition of a human endometrial carcinoma cell line xenografted into nude mice. Gynecol. Oncol. 17:124.

Fuller A.F., Guy S., Budzik G.P. and Donahoe P.K. 1982. Mullerian Inhibiting Substance inhibits colony growth of a human ovarian cell line. J. Clin. Endocrinol. Metab. 54:1051.

Fuller A.F., Krane I.M., Budzik G.P. and Donahoe P.K. 1985. Mullerian Inhibiting Substance reduction of colony growth of human. Gynecol. Oncol. 22:135.

Hayashi M., Shima H., Trelstad R. and Donahoe P.K. 1984. Immunocytochemical localization of Mullerian Inhibiting Substance in rough endoplasmic reticulum and Golgi apparatus in Sertoli cells of the neonatal calf using a monoclonal antibody. J. Histochem. Cytochem. 32:649.

Hutson J.M. and Donahoe P.K. 1986. The hormonal control of testicular descent. Endocrine Rev. 7:270.

Hutson J.M., Donahoe P.K. and MacLaughlin D.T. 1985. Steroid modualtion of Mullerian duct regeneration in the chick embryo. Gen. Comp. Endocrinol. 57:88.

Hutson J. M., Fallat M.E., Kamagata S., Donahoe P.K. and Budzik G.P. 1984. Phosphorylation events during Mullerian duct regression. Science 223:586.

Ikawa H., Hutson J.M., Budzik G.P., MacLaughlin D.T. and Donahoe P.K. 1982. Steroid enhancement of Mullerian duct regression. J. Ped. Surg. 17:453.

Josso N., Picard J.Y. and Tran D. 1977. The anti-Mullerian hormone. Recent Progm. Horm. Res. 33:117.

Jost A. 1946. Sur la differenciation sexuelle de l'embryon de lapin. Experiences de parabiose. C.R. Soc. Biol. 140:463.

Jost A. 1946. Sur la differenciation sexuelle de l'embryon de lapin remarques au sujet des certaines operations chirurgical sur l'embryon. C.R. Soc. Biol. 140:461.

Jost A. 1947. Sur les derives Mulleriens d'embryons de lapin des deux sexes castres a 21 jours. C.R. Soc. Biol. 141:135.

Necklaws E., LaQuaglia M., MacLaughlin D.T., Hudson P., Mudgett-Hunter M. and Donahoe P.K. 1986. Detection of Mullerian Inhibiting Substance in biological samples by a solid phase sandwich radioimmunoassay. Endocrinology 2:791.

Pepinsky R.B., Sinclair L.K., Chow E.P., Mattaliano R.J., Manganaro T.F., Donahoe P.K. and Cate R.L. 1988. Proteolytic processing of Mullerian Inhibiting Substance produces a TGF-β like fragment. J. Biol. Chem. 263:18961.

Picard J-Y., Benarous R., Guerrier D., Josso N., and Kahn A. 1986. Cloning and expression of cDNA for anti-Mullerian hormone. Biochemistry 83:5464.

Picon R. 1969. Action du testicule feotal sur le development *in vitro* des canaux de Muller chez le rat. Arch. Anat. Morph. Exp. 58:1.

Takahashi M., Hayashi M., Manganaro T.F., and Donahoe P.K. 1986a. The ontogeny of Mullerian Inhibiting Substance in granulosa cells of the bovine ovarian follicle. Biol. Reprod. 35:447.

Takahashi M., Koide S.S. and Donahoe P.K. 1986b. Mullerian Inhibiting Substance as oocyte meiosis inhibitor. Mol. Cell. Endocrinol. 47:225.
Ueno S., Manganaro T.F. and Donahoe P.K. 1988. Human recombinant Mullerian Inhibiting Substance inhibition of rat oocyte meiosis is reversed by epidermal growth *in vitro*. Encodrinol. 123:1652.

Vigier B., Picard J-Y., Tran D., Leageal L., and Josso N. 1984. Production of anti-Mullerian hormone: another homology between Sertoli and granulosa cells. Endocrinol. 114:1315.

Yamada T. 1974. The cellular biology of a newly established cell line of human ovarian adenocarcinoma *in vitro*. Keio J. Med. 23:53.

GROWTH FACTOR, ONCOGENE, AND STEROIDAL INTERACTIONS IN THE REGULATION OF UTERINE GROWTH AND FUNCTION

D.S. Loose-Mitchell*, C. Chiappetta*, R.M. Gardner*, J.L. Kirkland+, T.-H. Lin+, R.B. Lingham*, V.R. Mukku*, C. Orengo*, and G.M. Stancel*

*Department of Pharmacology
The University of Texas Medical School
Houston, TX 77225

+Division of Endocrinology
Department of Pediatrics
Baylor College of Medicine
Houston, TX 77030.

Historically, most studies of uterine growth and function in the non-pregnant uterus focused on the hormones estradiol and progesterone. In some cases, other hormones such as growth hormone (Grattarola and Li, 1959), triclothymine (T_3, Gardner *et al.*, 1978: Kirkland *et al.*, 1981 a) insulin (Kirkland *et al.*, 1981 b; Fowler *et al.*, 1963) and glucocorticoids (Szego and Roberts, 1953; Campbell, 1978) were also studied, but the principal emphasis remained on the two ovarian hormones noted above. These studies emphasized the structure, function, and regulation of steroid receptors; nuclear interactions of receptors; and biological responses resulting from hormonal stimulation of the tissue (Jensen and DeSombre, 1972; Gorski and Gannon, 1976). During this time, scant attention was given to possible interactions between steroids and polypeptide growth factors in the regulation of uterine growth and function.

By the 1980's, a number of peptide growth factors had been identified. These molecules support the growth of cultured cells, such as fibroblasts, in the absence of serum or in minimal levels of serum. In some cases these peptides are transported by serum or formed elements of the bloodstream (e.g., platelets) and in others they are produced and act "locally" via autocrine or paracrine mechanisms (Deuel, 1987). During this same time period, virologists were increasingly successful in identifying specific retroviral oncogenes (Bishop, 1985). These two streams of research began to overlap when it was realized that nontransformed cells contained analogs (so called protooncogenes or cellular oncogenes) of viral oncogenes; that some of these protooncogenes were either growth factors or growth factor receptors; and that some growth factors stimulated protooncogene expression (Deuel, 1987; Bishop, 1985). These findings suggested that protooncogenes and growth factors play an important role in the growth and function of normal cells.

Given these developments in the area of growth control, it was probably inevitable that laboratories studying steroid mediated growth would begin to investigate possible interactions between these hormones, peptide growth factors, and protooncogenes. This approach has clearly emerged within the last five years, and it has already been demonstrated that estrogen, androgen, progesterone and glucocorticoid interactions may involve growth factors and/or protooncogenes in a variety of systems. This literature is already so large as to preclude a thorough review in this chapter.

In the case of the uterus a number of growth factors and proto-oncogenes have already been identified, and we expect this list will continue to grow. Epidermal growth factor (EGF; Gonzalez *et al.*, 1984; DiAugustine *et al.*, 1988), insulin-like growth factor 1 (IGF-1; Murphy *et al.*, 1987 a; Murphy and Friesen, 1988), and a uterine derived growth factor (UDGF; Ikeda and Sirbasku, 1984) have been identified in the uterus; uterine luminal fluid contains EGF (Imai, 1982) and a more recently observed uterine luminal fluid mitogen (ULFM; Simmen *et al.*, 1988); and the uterus contains receptors for EGF (Hoffman *et al.*, 1984; Mukku and Stancel, 1985 a, b) and platelet-derived growth factor (PDGF; Ronnstrand *et al.*, 1987). Protooncogenes expressed in the uterus include N-myc (Murphy *et al.*, 1987 b), c-myc (Murphy *et al.*, 1987 b; Travers and Knowler, 1987), c-fos (Loose-Mitchell *et al.*, 1988; Weisz and Bresciani, 1988), c-ras ^{Ha} (Travers and Knowler, 1987), and c-erb B, i.e., the EGF receptor (Mukku and Stancel, 1985 a, b; Lingham *et al.*, 1988). In all cases estrogen treatment *in vivo* increases the mRNA levels of these protooncogenes. The uterus thus represents an organ well suited to study the possible role of growth factors and protooncogenes in the estrogen regulated growth of normal tissue *in vivo*.

In the next section of this chapter we will summarize findings from our laboratories on the regulation of c-fos and the EGF receptor in the uterus by estrogen, and a possible role for EGF in the control of myometrial contractility. In the subsequent section we will discuss our findings and the work of others and suggest an overall model for the interactions of estrogen, protooncogenes and growth factors in the control of uterine growth.

REGULATION OF C-FOS BY ESTROGEN

c-fos is the cellular analog of the v-fos oncogene identified originally in a retrovirus obtained from a murine osteosarcoma. The c-fos gene codes for a protein of 55,000 MW, but the protein is generally detected as a higher molecular weight form (approximately 62,000 MW) due to extensive posttranslational modifications. The c-fos protein is localized in the nucleus and is capable of binding DNA (Alt *et al.*, 1987), although its role in transcriptional regulation may be as part of a complex with other transcriptional activators such as AP-1/c-jun (Ruther *et al.*, 1987; Lech *et al.*, 1988; Distel *et al.*, 1987; Rauscher *et al.*, 1988; Chiu *et al.*, 1988). The expression of c-fos is rapidly and dramatically increased following stimulation of many cells with serum, specific growth factors, or a variety of other agents (Alt *et al.*, 1987). In addition, "anti-sense" experiments have indicated that c-fos is required for mitogenesis in 3T3 cells (Holt *et al.*, 1986; Nisikura and Murray, 1987). Given the possible role of c-fos as a transcriptional activator and the widespread activation of c-fos expression, we investigated the effect of estrogen treatment on c-fos expression *in vivo*.

For these studies, immature female rats were treated with estradiol, uteri were removed at sacrifice, and total RNA was prepared by the

HORMONAL SPECIFICITY

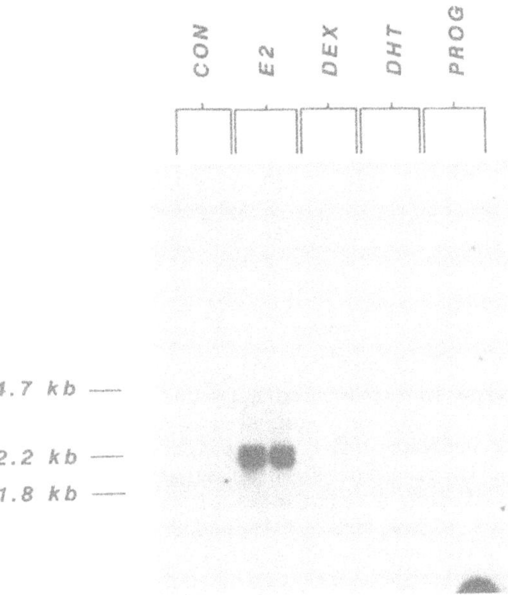

Figure 1. Induction of c-fos mRNA by estrogen. Immature female rats were treated with saline (CON), 40 μg/kg of estradiol (E_2), 600 μg/kg of dexamethasome (DEX), 400 μg/kg of dihydrotestosterone (DHT), or 40 mg/kg of progesterone (PROG) for 3 hours. Total uterine RNA was prepared and subsequently hybridized to c-fos antisense RNA. (Loose-Mitchell *et al.*, 1988, with permission).

guanidinium thiocyanate/cesium chloride method (Chirgwin *et al.*, 1979). Following agarose gel electrophoresis and transfer to nitrocellulose, ^{32}P-labelled cRNA probes were used to detect the 2.2 kb c-fos transcript. As seen in Figure 1, estradiol but not non-estrogenic steroids produced a massive increase in the level of c-fos mRNA.

Time course studies revealed that this increase in c-fos mRNA levels was very rapid, doubling within 30 minutes and reaching peak values (30-40 fold over controls) within 3 hours (Table 1). This rapid increase in c-fos mRNA levels is transient and rapidly declines, as has been observed in other systems. We have also shown that the increase in c-fos expression is sensitive to actinomycin D but not puromycin, and that induction occurs with doses of estrogen that stimulate uterine DNA synthesis (Loose-Mitchell *et al.*, 1988). Other workers have also observed this rapid increase in c-fos expression using nuclear run-on assays (Weisz and Bresciani, 1988).

Ovariectomized rats were treated with E_2 (40 μg/kg) and sacrificed at the indicated times. RNA was prepared and blot analysis performed as in Figure 1. The 2.2 kB c-fos bands of the resultant films were scanned with a laser densitometer. c-fos mRNA levels are expressed in relative units based upon control samples present on the same film. Values represent means with the indicated SEM. N represents the number of different samples analyzed at each time point. (Loose-Mitchell *et al.*, 1988, with permission).

Table 1. Time Course of c-fos mRNA induction following estradiol administration.

Time (hours)	c-fos mRNA	n
0	1.00 ± .22	5
1/2	2.12 ± .07	2
1	7.02 ± 1.55	6
3	31.5 ± 8.53	6
6	8.64 ± 1.33	4
12	10.4 ± 0.51	2
15	2.64 ± 0.08	2
18	1.83 ± 0.10	2

The rapid decline in expression suggests a very tight control of c-fos mRNA levels during the estrogen stimulated growth response. This prompted us to ask whether c-fos expression could be re-elevated by estrogen after the initial hormonal stimulation. For these studies, animals were treated with a first dose of estradiol for 6 hours or longer, i.e., to allow the initial increase and decrease in mRNA levels. Animals were then challenged with a second dose of estradiol for an additional 3 hours prior to sacrifice. The timing and dose levels used for the estrogen injections were chosen to insure that an ample level of estrogen receptors was available at the time of the second injection.

The results of this study indicated that c-fos induction was refractory to estradiol stimulation at 6, 12 or 24 hours after the first hormone injection. c-fos expression could be re-stimulated, however, if the estrogen injections were spaced 48 hours apart (Orengo *et al.*, unpublished observations). While it may be fortuitous, it is interesting to note that uterine DNA synthesis reaches a peak 24 hours after estrogen treatment and requires 36-48 hours to return to unstimulated levels (Kaye *et al.*, 1972; Kirkland *et al.*, 1979). This may indicate that a mechanism exists to prevent c-fos expression from occurring at inappropriate times during the cell cycle, although we have no information at present about the nature, level or physiological significance of this putative mechanism.

While c-fos expression may be an important component of the growth response in some systems, it is clear that an increase in c-fos alone is not sufficient to stimulate mitogenesis. For example, PDGF stimulates c-fos expression in fibroblasts (Rollins and Stiles, 1988), but this growth factor alone does not stimulate DNA synthesis. Mitogenesis requires the presence of PDGF plus additional growth factors, e.g., EGF or IGF-1 (Stiles *et al.*, 1979; Okeefe and Pledger, 1983; Deuel, 1987). Since early studies did not observe an increase in uterine EGF prior to DNA synthesis (Gonzalez *et al.*, 1984), we considered the possibility that estrogens might affect the sensitivity of uterine cells to EGF by regulating the level of its cognate receptor.

REGULATION OF EGF RECEPTOR LEVEL BY ESTROGEN

Prior to studies on hormonal regulation, we characterized the uterine EGF receptor system. Uterine membranes contain a single class of specific, saturable EGF receptor binding sites. The uterine receptor has a MW of 170,000 and contains an EGF stimulated tyrosine kinase activity. The K_d for specific EGF binding is 0.5-1 nM, and this is unchanged by

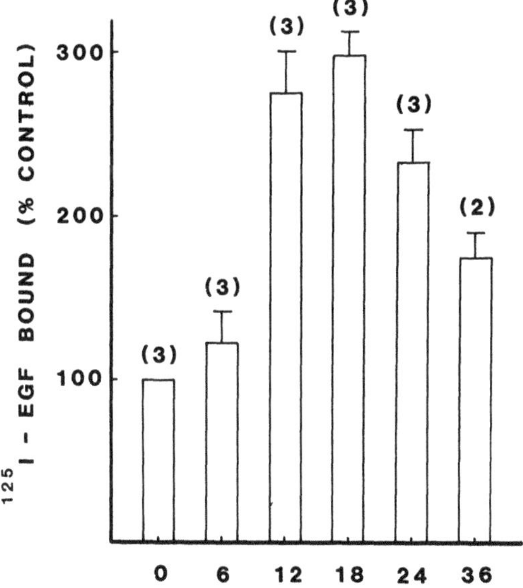

Figure 2. Induction of EGF receptors by estrogen. Immature female rats were treated with 40 μg/kg of estradiol for the indicated times prior to sacrifice and the preparation of uterine membranes. EGF receptor levels were measured by the specific binding of ^{125}I-EGF. (Mukku and Stancel, 1985a, with permission).

estrogen treatment *in vivo* (Mukku and Stancel, 1985a,b). The uterine EGF receptor is thus very similar to that found in a variety of normal and tumor cells (Carpenter, 1987). Autoradiographic studies in the rat (Lin *et al.*, 1988) and human (Chegini *et al.*, 1986) have shown that all major uterine cell types (luminal and glandular epithelium, stroma, and myometrium) contain EGF receptors, although receptor levels in various cell layers have not been quantitated precisely.

Treatment of immature rats with estradiol leads to an increase in functional EGF receptors in 6-12 hours as illustrated in Figure 2. This increase occurs well before the peak of tissue DNA synthesis seen 24 hours after estrogen administration (Kaye *et al.*, 1972; Mukku *et al.*, 1982). This increase in EGF receptor levels is specific for estrogenic steroids, involves a increase in the number of binding sites determined by Scatchard analysis and is accompanied by an increase in EGF-stimulated tyrosine kinase activity (Mukku and Stancel, 1985 a).

As illustrated in Figure 3, the increase in EGF receptor levels which occurs after estrogen treatment is sensitive to cycloheximide suggesting that the response represents *de novo* receptor synthesis. The increase is also sensitive to actinomycin D.

These experiments were conducted with immature rats (Mukku and Stancel, 1985 a), but similar results have also been obtained with mature, castrate rats (Gardner *et al.*, 1989), and with immature mice (unpublished observations). Studies with mature, cycling rats have also revealed that changes in uterine EGF receptor levels closely parallel

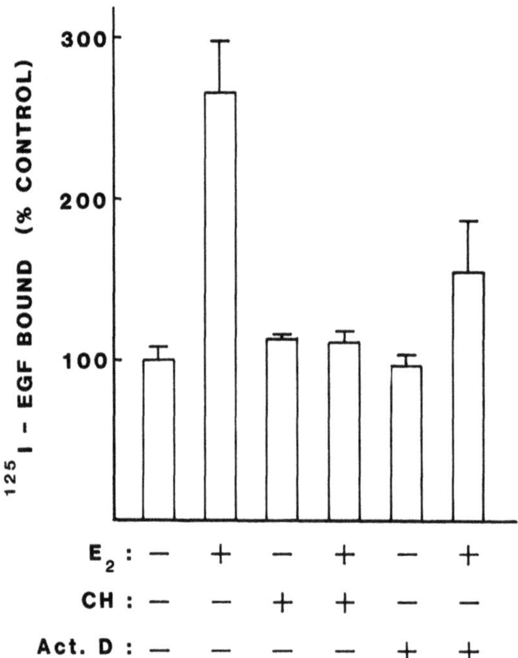

Figure 3. The effects of RNA and protein synthesis inhibitors on the induction of EGF receptors by estrogen. Immature female rats were treated with 40 μg/kg of estradiol (E_2), 4 mg/kg of cycloheximide (CH), or 8 mg/kg of actinomycin D (Act. D.), as indicated for 12 hours prior to sacrifice. Uterine membranes were prepared and assayed for EGF receptor levels by the specific binding of ^{125}I-EGF. (Mukku and Stancel, 1985a, with permission).

serum estrogen levels and values of occupied nuclear estrogen receptors throughout the estrous cycle (Gardner *et al.*, 1989). These results suggest that the regulation of EGF receptor levels by estrogen is a general physiologic phenomenon which occurs in different species and in different developmental states.

Because of the inhibitor studies shown in Figure 3, we next investigated EGF receptor mRNA levels after estrogen treatment. As seen in Figure 4, both the 9.5 and 6.6 kb EGF receptor transcripts increase after hormone administration and reach peak levels at 3-6 hours after injection.

Accompanying studies revealed that this increase in mRNA levels is: specific for estrogens; sensitive to actinomycin D, but not puromycin; and produced by doses of estradiol which stimulate tissue DNA synthesis (Lingham *et al.*, 1988). In parallel with these studies in the rat, we have also observed an increase in the abundance of EGF receptor mRNA after estrogen administration in the mouse (Gardner *et al.*, unpublished observations). Figure 5 is a composite profile illustrating the time course for the induction of EGF receptor mRNA and functional EGF receptors in the immature rat after hormone treatment.

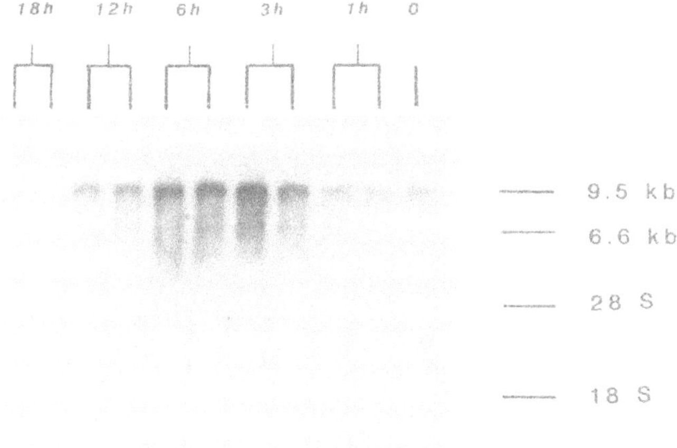

Figure 4. Induction of EGF receptor mRNA by estrogen. Immature female rats were treated with 40 μg/k of estradiol for the indicated times prior to sacrifice. Total uterine RNA was prepared and analyzed for EGF receptor transcript levels by blot analysis with antiense RNA. (Lingham *et al.*, 1988, with permission).

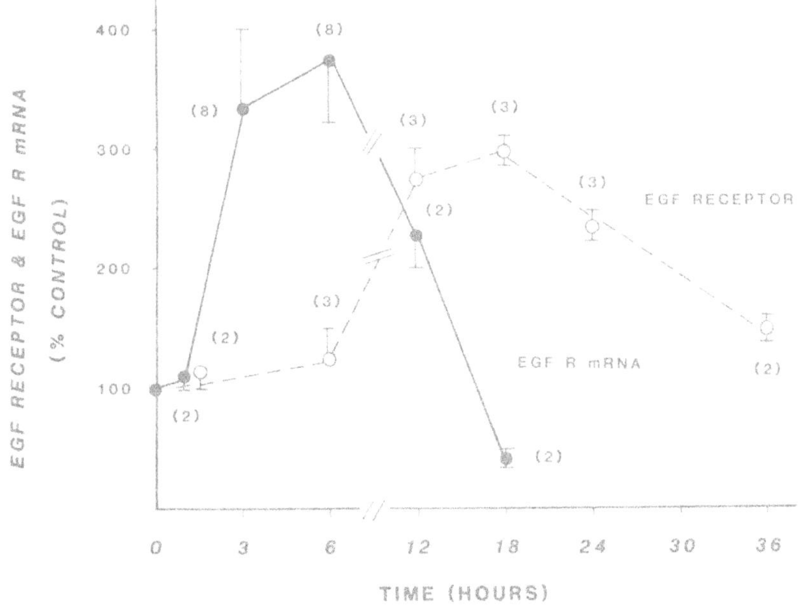

Figure 5. Composite of uterine EGF receptor and EGF receptor mRNA levels as a function of time after estrogen treatment. Animals were treated with 40 μg/kg of estradiol as indicated. Receptor levels (solid line) were measured by ^{125}I-EGF binding, and mRNA levels (dashed line) were measured by densitometric scanning of films from blots hybridized with antisense RNA. (Lingham *et al.*, 1988, with permission).

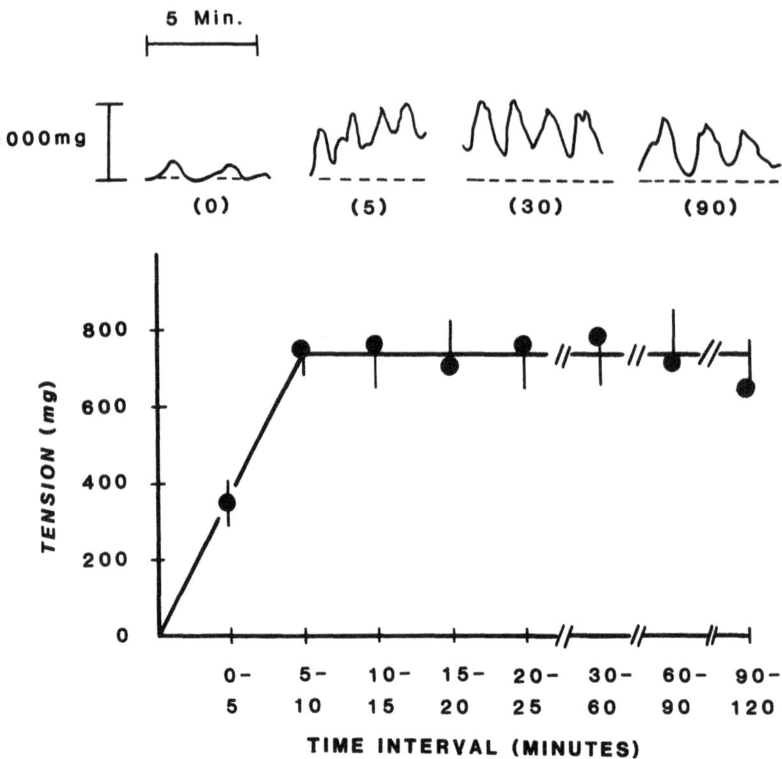

Figure 6. Stimulation of uterine contractions by EGF. Uterine strips from mature, ovariectomized rats were suspended *in vitro* and challenged with EGF (10^{-7}M). Top. Tracings from a single strip at the indicated times after EGF addition. Bottom. Maximum tension developed in a series of uterine tissues within the indicated time intervals after EGF addition (N=7 per point).

STIMULATION OF UTERINE CONTRACTILITY BY EGF

Our initial interest in EGF and its corresponding receptor was based on a possible role in uterine growth. However, several findings led us to consider the possibility that EGF might affect uterine function in other ways. These included the observations that EGF receptors are present in the myometrium as well as the endometrium (Chegini *et al.*, 1986; Lin *et al.*, 1988), EGF stimulates contractions of vascular smooth muscle (Berk *et al.*, 1985; Muramatsu *et al.*, 1985) and EGF increases prostaglandin production in certain cells (Shupnik and Tashjian, 1986; Yokota *et al.*, 1986; Chiba *et al.*, 1982). We thus investigated the effect of EGF on contractions of uterine smooth muscle.

For these studies, uterine strips were prepared, suspended in van Dykes - Hastings medium, and challenged with EGF. As seen in Figure 6, EGF addition caused a rapid and sustained increased in uterine contractility. This was evidenced by an increase in both overall muscle tone and rhythmicity.

To insure that the observed effect was due to EGF, we pre-incubated our EGF preparations with anti-EGF antibodies, and this abolished completely the increase in tension seen in response to the growth factor

(Gardner *et al.*, 1987). Other studies revealed that this effect of EGF is produced by EGF concentrations consistent with a receptor mediated effect; requires estrogen priming *in vivo* and occurs in tissues from both mature and immature rats (Gardner *et al.*, 1987). A contractile response to EGF is also observed with the mouse uterus (unpublished observations), indicating this effect is not limited to the rat.

More recent studies have shown that the contractile response of the myometrium is observed after the endometrium is removed, and that the effect is blocked by inhibitors of cyclooxgenase and lipoxygenase (Gardner *et al.*, 1988). This suggests that EGF produces uterine contractions by mobilization of myometrial arachidonic acid and subsequent production of metabolites (e.g., prostaglandins and leukotrienes) via cyclooxygenase and lipoxygenase catalyzed pathways.

These findings also serve to illustrate that peptides such as EGF may have diverse biological activities other than their roles as growth factors. We are currently pursuing studies of the contractile effect of EGF to more precisely determine its cellular basis and to assess its possible biological significance.

INTERACTIONS OF ESTROGEN, PROTOONCOGENES AND GROWTH FACTORS IN THE CONTROL OF UTERINE GROWTH

Estrogen Induction of c-fos Expression

We (Loose-Mitchell *et al.*, 1988) and others (Weisz and Bresciani, 1988) have shown that estrogen rapidly stimulates uterine c-fos expression *in vivo*. This regulation almost certainly occurs at the transcriptional level, at least in part, for a number of reasons. First, the increase in expression is very rapid and is blocked by inhibitors of transcription but not translation. Secondly, the upstream region of the c-fos gene contains the palindromic sequence 5′ GGTCT ... AGACC -3′ at -219/-207 (Treismann, 1985), which is very similar to the estrogen responsive element, ′5 -GGTCA ... TGACC -3′, identified in the Xenopus vitellogenin A_2 gene (Klein-Hitpass, 1986). Third, specificity and dose response studies are consistent with an estrogen receptor mediated effect. Finally, nuclear run-on measurements from Bresciani's group (Weisz and Bresciani, 1988) have demonstrated directly an increase in transcription.

While the mechanism of c-fos induction by estrogen seems reasonably well established, the role of fos in uterine growth can only be speculative at this time. Recent results on the role of fos in other systems, however, provide some exciting possibilities.

First, fos expression stimulates the transcription of other genes. This is thought to result from the formation of a complex between fos and other transcriptional regulators such as the AP-1/c-jun protein, with the complex itself being directly responsible for activation of target gene expression (Ruther *et al.*, 1987; Lech *et al.*, 1988; Distel *et al.*, 1987; Rauscher *et al.*, 1988; Chiu *et al.*, 1988). The fos /AP-1 complex could interact with numerous genes having the required AP-1 regulatory sequence. Furthermore, c-fos is known to interact with a number of other proteins, the so-called "fos associated proteins" or FAPs (Alt *et al.*, 1987). This would provide a mechanism for an initial signal from the estrogen receptor:estrogen responsive element (ERE) interaction (at the c-fos gene) to be amplified, since the fos/FAP complex could enhance transcription of other genes which do not contain ERE sequences. Such an amplification mechanism would also have a temporal aspect, i.e., the sharp "spike" of fos expression would regulate the times at which these "secondary" genes were expressed in the cell cycle. Such an amplification mechanism would seem well suited to play a role in the massive growth response of the uterus to estrogens.

Second, the c-fos gene itself is known to be regulated by a variety of agents other than estrogen, and regulatory sequences other than the ERE are utilized by these other agents to activate c-fos transcription. Similarly, the genes activated by complexes such as fos/AP-1 are likely to be subject to regulation by other stimuli, e.g., agents which activate protein kinase C (Ruther *et al.*, 1987; Lech *et al.*, 1988; Distel *et al.*, 1987; Rauscher *et al.*, 1988; Chiu *et al.*, 1988). These considerations provide two levels at which estrogen induced responses could be modulated by other stimuli, and vice versa. The first level is the c-fos gene itself, and the second level involves genes activated by fos-containing complexes. While these suggestions are admittedly speculative they provide a mechanism consistent with our current knowledge to allow "cross-talk" between estrogen and other signalling systems.

In addition to fos, the expression of other protooncogenes is activated following estrogenic stimulation of the uterus. These include N-myc (Murphy *et al.*, 1987 b), c-myc (Murphy *et al.*, 1987 b; Travers and Knowler, 1987), c-ras ^{Ha} (Travers and Knowler, 1987), and erb B (Mukku and Stancel, 1985 a; Lingham *et al.*, 1988). Thus, prototypes of all major protooncogene classes (i.e., nuclear protooncogenes, "G"-proteins, and protein kineses) are activated, but the mechanism of these effects is not established as clearly as that of fos induction. Further studies of fos and other oncogenes will hopefully provide additional insights into the physiological, pharmacological and pathological effects of estrogens on normal target tissue.

Estrogen Regulation of EGF Receptor Levels

Estrogens clearly regulate the level of functional EGF receptors in the uterus. On the basis of specific activity (i.e., content of membrane protein), estradiol causes a 3-fold increase in receptor levels. On a cellular basis (i.e., based on uterine DNA content) the peak increase is 4- to 5-fold since uterine protein content increases substantially prior to the onset of DNA synthesis. This increase occurs without alterations in receptor properties such as ligand affinity or molecular weight (Mukku and Stancel, 1985 a). While the mechanism of induction is not unequivocally established, inhibitor studies and the increase in EGF receptor mRNA content suggest that the regulation is at least partly transcriptional (Lingham *et al.*, 1988; Mukku and Stancel, 1985 a).

A major question at present is the role of EGF in uterine growth. Uterine tissue (Gonzalez *et al.*, 1984; DiAugustine *et al.*, 1988) and luminal fluid (Imai, 1982) contain EGF; the uterus contains prepro-EGF mRNA sequences (DiAugustine *et al.*, 1988); and EGF receptors are clearly present in the tissue. More importantly for this question, however, are a number of other observations. First, EGF stimulates the growth of cultured smooth muscle cells derived originally from the myometrium (Bhargava et al., 1979). Secondly, EGF stimulates the growth of cultured uterine epithelial cells from the mouse (Tomooka *et al.*, 1986). Finally, antibodies against EGF block estrogen-induced growth of uterine tissue segments in organ culture (McLachlan *et al.*, 1987). Taken together, these results provide substantial evidence for a role of EGF in the estrogen induced uterine growth *in vivo*.

Given these findings, regulation of EGF receptor levels could play several roles. First, an increase in receptor levels, alone or in combination with increases in tissue EGF levels, could be necessary to reach or maintain a threshold level of a second messenger necessary for cell growth and DNA synthesis. Second, during times of rapid tissue growth, receptor synthesis may be increased to offset down regulation and insure that the cellular content of receptors does not fall below a critical level required for growth. This would be consistent with previous findings in cultured fibroblasts that cells must be exposed to EGF for a considerable length of time (e.g., 8-10 hours) to obtain a maximum increase in DNA synthesis (Carpenter and Cohen, 1976), and that cell growth correlates with the number of EGF-receptor complexes present at steady-state after the initial down regulation seen following EGF addition to the media (Knauer *et al.*, 1984).

Another major question is how EGF might interact with other growth factors in the overall process of uterine growth. As noted previously, the uterus also contains IGF-1 (Murphy *et al.*, 1987 a; Murphy and Friesen, 1988), UDGF (Ikeda and Sirbasku, 1984), and ULFM (Simmen *et al.*, 1988), and PDGF receptors are also present in the tissue (Ronnstrand *et al.*, 1987). IGF-1 in particular seems a likely candidate for a role in uterine growth since uterine expression of this growth factor is increased by estrogen administration (Murphy *et al.*, 1987 a).

Stimulation of Uterine Contractions by EGF

The observation that EGF stimulates contractile activity of the uterus was an unexpected finding (Gardner *et al.*, 1987), but raises several important points. First, it provides the experimental ability to monitor the functional status of uterine EGF receptors, at least in the myometrium. Second, it raises the possibility that EGF plays a physiological role in the regulation of uterine contractility. Third, the tone of uterine smooth muscle is one factor that affects blood flow to the endometrium, and this parameter could conceivably affect uterine growth and function, e.g., by affecting the supply of nutrients and other serum factors. These possibilities are highly conjectural at this point, and much work remains to be done in order to evaluate rigorously these possibilities.

These findings do suggest, however, that we and others need to be cautious in suggesting that all effects of growth factors in the uterus are related to growth of the organ. It is entirely possible that EGF and other growth factors may have a number of diverse biological activities, some of which may be related more to differentiated function than growth.

Overall Uterine Growth in Response to Estrogen

It has been recognized for some time that the regulation of uterine growth by estrogens is a process which is regulated in several stages and requires more than an initial, transient interaction of the estrogen receptor with an agonist. For example, so called "short - acting" estrogens can interact with the receptor and produce rapid responses similar to those produced by estradiol. These compounds, however, cannot stimulate the complete uterine response culminating in DNA synthesis and cell division (Anderson *et al.*, 1972, 1973). Endocrinologists have referred to these "stages" of the estrogen induced growth response as "early" and "late effects", "late effects" also being referred to as "true uterine growth". More recently, we have been struck by the general similarity between this model of uterine growth and models of growth in cultured fibroblasts.

In the fibroblast model, the mitogenic stimulation of quiescent cells (in G_0) occurs in at least two major "steps". The first has been termed "competence" and is thought to represent the movement of arrested cells from G_0 into the early portion of G_1 or at least to the G_0/G_1 interface. Movement of cells through G_1 toward S is termed "progression". In the fibroblast system there is a clear demarcation between "competence" and "progression" because the two stages are controlled by different peptide factors. PDGF is the prototype "competence" factor, but alone cannot stimulate mitogenesis. Cells must first be exposed to PDGF and then to a "progression" factor such as EGF or IGF-1 (Stiles *et al.*, 1979; Deuel, 1987; O'Keefe and Pledger, 1983). In the uterine system, however, this distinction is not as sharp because estrogen itself seems to be involved in both "early" and "late" effects (Anderson *et al.*, 1972, 1973).

The analogy between the two systems is even more striking if one considers that PDGF regulates fos and myc expression in fibroblasts (Rollins and Stiles, 1988) and estradiol regulates fos (Weisz and Bresciani, 1988; Loose-Mitchell *et al.*, 1988) and myc (Travers and Knowler, 1987; Murphy *et al.*, 1987 b) expression in the uterus. While the two stimuli have in common the regulation of these specific genes, mechanistic steps between receptor-agonist interaction and gene expression are clearly different for PDGF and the steroid. Similarly, EGF and IGF-1 are "progression" factors for fibroblasts, and estradiol seems to increase the expression of IGF-1 (Murphy *et al.*, 1987 a), the level of EGF (DiAugustine *et al.*, 1988), and the level of EGF receptors (Mukku and Stancel, 1985 a; Lingham *et al.*, 1988) in the uterus. It thus seems reasonable to suggest that the so-called "early" and "late" phases of estrogen action may be analogous in a general way to "competence" and "progression" in fibroblasts.

If one considers fos expression as a marker of the competence state, estrogenic steroids may function via their receptors to elicit competence responses directly at the genomic level. Given our data and the work of others, this seems the most reasonable possibility. While it seems less likely to us, estrogens could conceivably act via a non-genomic mechanism (e.g., conversion of an inactive EGF precursor to an active peptide) which functions through a membrane receptor to stimulate c-fos expression. PDGF-like regulation, on the other hand, involves a cytoplasmic second messenger(s) to transmit a signal enamating from the plasma membrane. The basic idea is that steroids and polypeptide growth factors control "competence" by different regulatory mechanisms which converge at a common endpoint(s) such as expression of c-fos.

Possible mechanistic analogies between "progression-like" events in the two systems are more difficult to envision. An obvious possibility is that the estrogen-receptor complex directly controls the local production of factors like EGF/IGF-1 and/or their receptors' and these act via autocrine or paracrine mechanisms to control transit through G_1. Unfortunately, it is difficult to evaluate this possibility given existing data on the mechanism of regulation of EGF and IGF-1 by estrogens. Additional studies will obviously be required to evaluate this possibility, and to understand the *in vivo* role of protooncogenes and growth factors in estrogen controlled uterine growth.

ACKNOWLEDGEMENTS

The authors wish to express their appreciation to Ms. Martha Zurita for preparation of this manuscript. Work from our laboratories was supported by NIH grants HD-08615 (G.M.S), DK-38965 (D.S.L.-M.), and RR-01685 (Bionet) for computer use, and a grant from the John P. McGovern Foundation (J.L.K.).

REFERENCES

Alt F.W., Harlow E., and Ziff E.B. 1987. In: "Nuclear oncogenes". Cold Spring Harbor Laboratory, Cold Spring Harbor, N.Y.

Anderson J.N., Clark J.H., and Peck E.J., Jr. 1972. The relationship between nuclear receptor estrogen binding and uterotrophic responses. Biochem. Biophys. Res. Commun. 48:1460.

Anderson J.N., Peck E.J., Jr., and Clark J.H. 1973. Nuclear receptor estrogen complex: relationship between concentration and early uterotrophic responses. Endocrinology. 92:1488.

Berk B.C., Brock T.A., Webb R.C., Taubman M.B., Atkinson W.J., Gimbrone M.A., Jr., and Alexander R.W. 1985. Epidermal growth factor, a vascular smooth muscle mitogen, induces rat aortic contraction. J. Clin. Invest. 75:1083.

Bhargava G., Rifas L., and Makman M.H. 1979. Presence of epidermal growth factor receptors and influence of epidermal growth factor on proliferation and aging in cultured smooth muscle cells. J. Cell. Physiol. 100:365.

Bishop J.M. 1985. Viral Oncogenes. Cell 42:23.

Campbell P.S. 1978. The mechanism of inhibition of uterotrophic responses by acute dexamethasone treatment. Endocrinology 103:716.

Carpenter G. 1987. Receptors for epidermal growth factor and other polypeptide mitogens. Ann. Rev. Biochemistry 56:881.

Carpenter G., and Cohen S. 1976. ^{125}I-labeled epidermal growth factor: binding, internalization and degradation in human fibroblasts. J. Cell Biol. 71:159.

Chegini N., Rao C.V., Wakin N., and Sanfilippo J. 1986. Binding of ^{125}I-epidermal growth factor in human uterus. Cell Tissue Res. 246:543.

Chiba T., Hirata T., Taminato T., Kadowaki S., Matsukura S., and Fujita T. 1982. Epidermal growth factor stimulate prostaglandin E release from isolated perfused rat stomach. Biochem. Biophys. Res. Commun. 105:370.

Chirgwin J.M., Przybyla A.E., MacDonald R.J., and Rutter W.J. 1979. Isolation of biologically active ribonucleic acids from sources enriched in ribonuclease. Biochemistry 18:5294.

Chiu R., Boyle W.J., Meek J., Smeol T., Hunter T., and Karin M. 1988. The c-fos protein interacts with c-Jun/AP-1 to stimulate transcription of AP-1 responsive genes. Cell 54:541.

Deuel T.F. 1987. Polypeptide growth factors: roles in normal and abnormal cell growth. Ann. Rev. Cell Biology 3:443.

DiAugustine R.P., Petrusz P., Bell G.I., Brown C.F., Korach K.S., McLachlan J.A., and Teng C.T. 1988. Influence of estrogen on mouse uterine epidermal growth factor precursor protein and messenger ribonucleic acid. Endocrinology 122:2355.

Distel R.J., Ro H.S., Rosen B.S., Groves D.L., and Spiegelman B.M. 1987. Nucleoprotein complexes that regulate gene expression in adipocyte differentiation: Direct participation of c-fos. Cell 49:835.

Fowler D.D., Szego C.M., and Sloan S.H. 1963. Curtailment of the uterotrophic action of estrogen by impaired histamine liberation in the alloxan-diabetic rat: reversal by insulin and by adrenalectomy. Endocrinology 72:626.

Gardner R.M., Kirkland J.L., Ireland J.S., and Stancel G.M. 1978. Regulation of the uterine response to estrogen by thyroid hormone. Endocrinology 103:1164.

Gardner R.M., Lingham R.B., and Stancel G.M. 1987. Epidermal growth factor stimulates contractions of the isolated uterus. FASEB J. 1:224.

Gardner R.M., Goldsmith J.R., and Stancel G.M. 1988. Pharmacological characterization of epidermal growth factor (EGF) induced uterine contractions. FASEB J. 2:A1143.

Gardner R.M., Verner G., Kirkland J.L., and Stancel G.M. 1989. Regulation of EGF receptors by estrogen in the mature rat and during the estrous cycle. J. Steroid Biochem. (in press).

Gonzalez F., Lakshmanan J., Hoath S., and Fisher D.A. 1984. Effect of oestradiol-17β on uterine epidermal growth factor concentration in immature mice. Acta Endocrinol. (Copenhagen) 105:425.

Gorski J. and Gannon F. 1976. Current models of steroid hormone action, a critique. Ann. Rev. Physiology 38:425.

Grattarola R., and Li C.H. 1959. Effect of growth hormone and its combination with estradiol on the uterus of hypophysectomized and hypophysectomized-ovariectomized rats. Endocrinology 65:802.

Hoffman G.E., Rao C.V., Barrows G.H., and Sanfilippo J.S. 1984. Binding sites for epidermal growth factor in human uterine tissues and leiomyomas. J. Clin. Endocrinol. Metab. 58:880.

Holt J.T., Gopal T.V., Moulton A.D., and Nienhuis A.W. 1986. Inducible production of c-fos antisense RNA inhibits 3T3 cell proliferation. Proc. Natl. Acad. Sci. (USA) 83:4794.

Ikeda T., and Sirbasku P.A. 1984. Purification and properties of a mammary uterine - pituitary tumor cell growth factor from pregnant sheep uterus. J. Biol. Chem. 259:4049.

Imai Y. 1982. Epidermal growth factor in rat uterine luminal fluid. Endocrinology 110 (Suppl): 162 (Abstract).

Jensen E.V., and DeSombre E.R. 1972. Mechanism of action of female sex hormones. Annu. Rev. Biochemistry 41:203.

Kaye A.M., Sheratzky D., and Lindner H.R. 1972. Kinetics of DNA synthesis in immature rat uterus: age dependence and estradiol stimulation. Biochim. Biophys. Acta 261:475.

Kirkland J.L., LaPointe L.B., Justin E., and Stancel G.M. 1979. The effect of estrogen treatment of mitosis in individual uterine cell types. Biol. Reprod. 21:269.

Kirkland J.L., Gardner R.M., Mukku V.R., Aktar M, and Stancel G.M. 1981a. Hormonal Control of Uterine Growth: The effect of hypothyroidism on estrogen stimulated cell division. Endocrinology 108:2346.

Kirkland J.L., Barrett G.N., and Stancel G.M. 1981b. Decreased cell division of the uterine luminal epithelium of diabetic rats in response to estradiol. Endocrinology 109:316.

Klein-Hitpass L., Schorp M., Wagner U., and Ryfell G.U. 1986. An estrogen responsive element derived from the 5' flanking region of Xenopus vitellogenin A2 gene functions in transfected human cells. Cell 46:1053.

Knauer D.J., Wiley H.S., and Cunningham D.B. 1984. Relationship between epidermal growth factor receptor occupancy and mitogenic response. Quantitative analysis using a steady state model system. J. Biol. Chem. 259:5623.

Lech K., Anderson K. and Brent R. 1988. DNA-bound c-fos proteins activate transcription in yeast. Cell 52:179.

Lin T.S., Kirkland J.L., Mukku V.R., and Stancel G.M. 1988. Autoradiographic localization of epidermal growth factor binding in individual uterine cell types. Biol. Reprod. 38:403.

Lingham R.B., Stancel G.M. and Loose-Mitchell D.M. 1988. Regulation of epidermal growth factor receptor messenger RNA by estrogen. Mol. Endocrinol. 2:230.

Loose-Mitchell D.S., Chiappetta C., and Stancel G.M. 1988. Estrogen regulation c-fos messenger ribonucleic acid. Mol. Endocrinol. 2:946.

McLachlan J.A., DiAugustine R.P., and Newbold R.R. 1987. Estrogen induced uterine cell proliferation in organ culture is inhibited by antibodies to epidermal growth factor. Program of the 69th Annual Meeting of the Endocrine Society, Indianapolis, IN, p. 99 (Abstract #313).

Mukku V.R., Kirkland J.L., Hardy M., and Stancel G.M. 1982. Hormonal control of uterine growth: temporal relationships between estrogen administration and DNA synthesis. Endocrinology 111:480.

Mukku V.R., and Stancel G.M. 1985a. Regulation of uterine epidermal growth factor receptors by estrogen. J. Biol. Chem. 260:9820.

Mukku V.R., and Stancel G.M. 1985b. Receptors for epidermal growth factor in the rat uterus. Endocrinology 117:149.

Muramatsu I., Hollenberg M.D., and Lederis K. 1985. Vascular actions of epidermal growth factor-urogastrone: possible relationship to prostaglandin production. Can. J. Physiol. Pharmacol. 63:994.

Murphy L.J., Murphy L.C., and Friesen H.G. 1987a. Estrogen induces insulin-like growth factor-1 expression in the rat uterus. Mol. Endocrinol. 1:445.

Murphy L.J., Murphy L.C., and Friesen H.G. 1987b. Estrogen induction of N-myc and c-myc protooncogene expression in the rat uterus. Endocrinology 120:1882.

Murphy L.J., and Friesen H.G. 1988. Differential effects of estrogen and growth hormone on uterine and hepatic insulin-like growth factor I gene expression in the ovariectomized hypophysectomized rat. Endocrinology 122:325.

Nishikura K., and Murray J.M. 1987. Antisense RNA of c-fos blocks renewed growth of quiesent 3T3 cells. Molec. Cell. Biol. 7:636.

O'Keefe E.J., and Pledger W.J. 1983. A model of cell cycle control of sequential events regulated by growth factors. Mol. Cell Endocrinol. 31:167.

Rauscher F.J., III, Sambucetti L.C., Curran T., Distel R.J., and Speigleman B.M. 1988. A common DNA binding site for fos protein and transcription factor AP-1. Cell 52:471.

Rollins B.J., and Stiles C.D. 1988. Regulation of c-myc and c-fos proto-oncogene expression by animal cell growth factors, in vitro. Cell. Develop. Biol. 24:81.

Ronnstrand L., Beckmann M.P., Faulders B., Ostman A., Ek B., and Heldin C.H. 1987. Purification of the receptor for PDGF from porcine uterus. J. Biol. Chem. 262:2929.

Ruther U., Garber C., Komitowski D., Muller R., and Wagner E.F. 1987. Deregulated c-fos expression interferes with normal bone development in transgenic mice. Nature (London) 325:412.

Shupnik M.A., and Tashjian A.H., Jr. 1982. Epidermal growth factor and phorbol ester actions on human osteosarcoma cells. Characterization of responsive and non responsive cell lines. J. Biol. Chem. 257:12161.

Simmen R.C.M., Ko Y., Liu X.H., Wilde M.H., Pope W.F., and Simmen F.A. 1988. A uterine cell mitogen distinct from epidermal growth factor in porcine uterine luminal fluids: characterization and partial purification. Biol. Reprod. 38:551.

Stiles C.D., Capone G.T., Scher C.D., Antoniades H.N., Van Wyk J.J., and Pledger W.J. 1979. Dual control of cell growth by somatomedins and platelet-derived growth factor. Proc. Natl. Acad. Sci. (USA) 76:1279.

Szego C.M. and Roberts S. 1953. Steroid actions and interaction in uterine metabolism. Recent Prog. Horm. Res. 8:419.

Tomooka Y., DiAugustine R.P., and McLachlan J.A. 1986. Proliferation of mouse uterine epithelial cells in vitro. Endocrinology 118:1011.

Travers M.T., and Knowler J.T. 1987. Oestrogen - induced expression of oncogenes in the immature rat uterus. FEBS Lett. 211:27.

Treisman R. 1985. Transient accumulation of c-fos RNA following serum stimulation requires a conserved 5' element and c-fos 3' sequences. Cell 42:889.

Weisz A., and Bresciani F. 1988. Estrogen induces expression of c-fos and c-myc protooncogenes in rat uterus. Mol. Endocrinol. 2:816.

Yokota K., Kusaka M., Ohshima T., Yamamoto S., Kurihara N., Yoshino T., and Kumegawa M. 1985. Stimulation of prostaglandin E-2 synthesis in cloned osteoblastic cells of mouse (MC 3T3-E1) by epidermal growth factor. J. Biol. Chem. 261:15410.

STEROID-INDUCED PROTEINS OF THE PRIMATE OVIDUCT AND UTERUS: POTENTIAL REGULATORS OF REPRODUCTIVE FUNCTION

Asgerally T. Fazleabas*
Harold G. Verhage* and
Stephen C. Bell+

*Department of Obstetrics and Gynecology
University of Illinois College of Medicine
Chicago, Illinois, U.S.A.

+Departments of Obstetrics and Gynecology
and Biochemistry
University of Leicester
Leicester, U.K.

The biological requirements of the mammalian conceptus during the initial stages of development must be met by oviductal and uterine secretions since they constitute the primary environmental contact between the developing embryo and its mother prior to implantation. Both the mammalian oviduct and uterine endometrium are target tissues for the ovarian steroids and each responds by undergoing cyclic changes in morphology and secretory activity. These cyclic changes prepare the oviduct and uterus to receive and nourish the gametes during fertilization, early embryonic development and during the implantation process. Secretory products from these two tissue compartments are necessary to maintain gamete viability, however, their specific role in primate reproduction has yet to be determined. Studies to determine critical reproductive events during these very early stages of embryonic development and pregnancy obviously are very difficult, if not impossible, to perform in humans. Therefore, we have attempted to utilize the baboon (*Papio anubis*) as a non-human primate model to identify hormonally regulated secretory proteins of the oviduct and uterus and to determine similarities and dissimilarities in the secretory profile of these two tissue compartments in the baboon and human. This chapter summarizes our findings which suggests potential paracrine roles for specific, hormonally regulated proteins of the baboon and human oviduct and uterus.

Studies on the baboon oviduct and uterus have characterized the morphological and secretory changes associated with specific stages of the menstrual cycle, namely early and late follicular and mid and late luteal. In order to confirm that the cyclic morphological and secretory changes are controlled by the ovarian steroids, we have also utilized an ovariectomized, steroid treated model that is summarized in Figure 1 and that has been described in detail elsewhere (Fazleabas *et al.*, 1988; Verhage and Fazleabas, 1988).

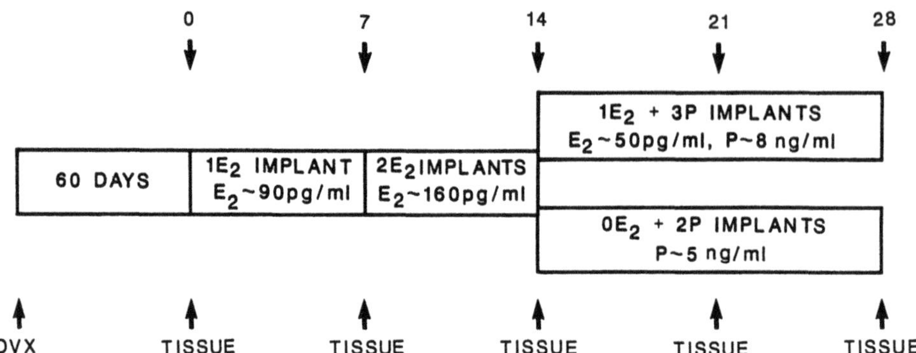

Figure 1. Experimental design for steroid treatment. Silastic capsules (6cm) filled with either 17β-estradiol (E_2) or progesterone (P) were inserted sc in the midcapsulary region of the ovariectomized baboon. E_2 was administered for either 7 days (one E_2 implant) or 14 days (two E_2 implants for the last 7 days). All P-treated animals were primed with this 14-day E_2 protocol before administering P in the presence or absence of E_2 (Verhage and Fazleabas, 1988, reprinted with permission).

OVIDUCT

Morphology

The primate oviduct is a target tissue for the ovarian steroids (see Verhage and Jaffe, 1986 and Brenner and Maslar, 1988 for reviews) and the epithelial cells lining the lumen appear to be the most sensitive tissue compartment to fluctuating steroid levels. At the time of ovulation in the cycling primate or during periods of estrogen dominance in ovariectomized animals, the oviductal luminal epithelium consists of columnar ciliated and secretory cells in approximately equal proportions. However, during periods of progesterone dominance the epithelium is made up primarily of cuboidal cells which are neither ciliated nor secretory.

The morphological changes observed during the cycle were compared to those observed in our steroid treated animals (Figure 2). The lining epithelium in the ampulla of the estradiol-dominated baboon consists of columnar ciliated and secretory cells, similar to the epithelium in humans (Verhage *et al.*, 1979) and other nonhuman primates (Brenner and Maslar, 1988; Verhage *et al.*, 1989) at the time of ovulation. The mature secretory cells lining the lumen of the oviduct have conspicuous apical blebs which contain secretory granules. In contrast, the apical blebs of the secretory cells found in the crypts of the longitudinal mucosal folds are less distinct, and the distribution of secretory granules are more general throughout the supranuclear cytoplasm. Electron microscopic analyses show that the apical tips of mature secretory cells in an estrogen-dominated primate oviduct contains secretory granules and that the release of some of this product occurs by the process of exocytosis (Odor *et al.*, 1983; Verhage and Jaffe, 1986). The epithelium lining the oviduct in the ovariectomized animals is comprised of a single cuboidal cell type which is neither ciliated nor secretory. Treatment with

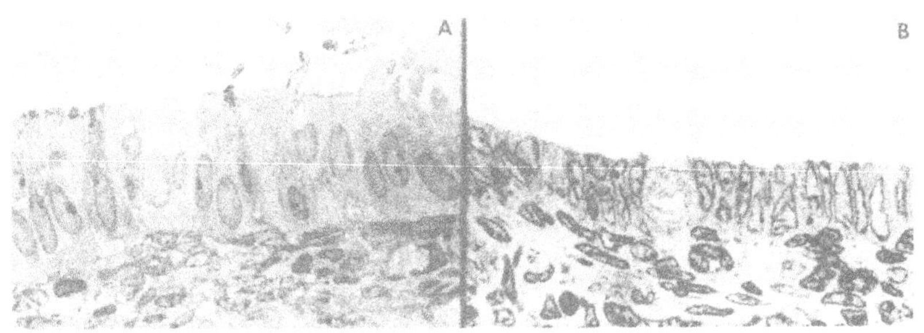

Figure 2. One micron plastic sections obtained from regions lining the lumen of the baboon ampulla stained with toluidine blue. Panel A, 14 day estradiol-treated animal. The epithelium is tall-columnar consisting of mature ciliated and secretory cells. The apical tips f the secretory cells form apical blebs and most of the basophilic secretory granules are found in these apical blebs. Panel B, estradiol-primed animal treated for 14 days with estradiol plus progesterone. The epithelium is simple cuboidal, similar to that observed in ovariectomized animals. An occasional ciliated cell is still present. Magnification x540.

progesterone following estrogen priming shows a morphological pattern similar to that observed in ovariectomized animals (Figure 2). The lining epithelium primarily consists of cuboidal cells which are neither ciliated nor secretory, similar to what is observed in the cycling animal just prior to menses. Occasionally, a few basophilic granules are still observed in some cells and a few still have cilia.

Secretory Proteins of the Baboon Oviduct

Oviduct fluid provides the optimal environment for fertilization and early embryonic development. We have utilized an explant culture system to determine the pattern of proteins synthesized by the baboon oviduct during the menstrual cycle (Fazleabas and Verhage, 1986) and in steroid-treated ovariectomized animals (Verhage and Fazleabas, 1988). A group of glycoproteins (M_r 100,000-130,000) are synthesized during the follicular stage of the menstrual cycle and in estrogen treated ovariectomized baboons (Figures 3 and 5A and B). These macromolecules comprise a heterogeneous group of glycoproteins of varying molecular weight (between 100,000 and 130,000) and pI (>8.0 to 4.5). An electrophoretically similar group of proteins are also synthesized by the human oviduct at mid-cycle (Verhage *et al.*, 1988; Figures 4 and 5C and D). Progesterone, either during the luteal stage of the menstrual cycle or when added to the steroid treatment regimen, essentially inhibits the synthesis of this group of glycoproteins (Figures 3 and 4). These data suggest that the synthesis and release of oviduct-specific macromolecules is cyclic in nature, perhaps dependent on the presence of estradiol and the absence of progesterone. Besides being synthesized *in vitro*, the oviduct-specific macromolecules are secreted into the oviductal lumen during the follicular stage of the cycle and with estrogen treatment, and contribute significantly to the oviductal fluid milieu (Fazleabas and Verhage, 1986; Verhage *et al.*, 1989).

The profile of protein synthetic activity observed in our studies supports the concept first proposed by Brenner *et al.* (1979) that progesterone is an off-switch for the primate oviduct, fully capable of antagonizing the action of estradiol. This action of progesterone is

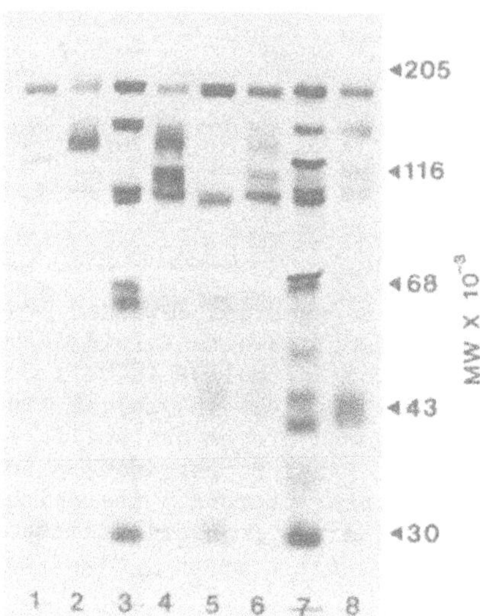

Figure 3. Fluorographs of one dimensional gels (7.5% SDS-PAGE) of baboon oviduct culture medium labelled with L-[^{3}H] leucine (odd-numbered lanes) or D-[6-^{3}H] glucosamine (even-numbered lanes). Tissue was obtained from the early follicular (lanes 1 and 2), late follicular (lanes 3 and 4), midluteal (lanes 5 and 6) and late luteal (lanes 7 and 8) stages of the menstrual cycle. Note the intensification of labeled secretory products (M_r 100,000-130,000) during the late follicular stage (lanes 3 and 4).

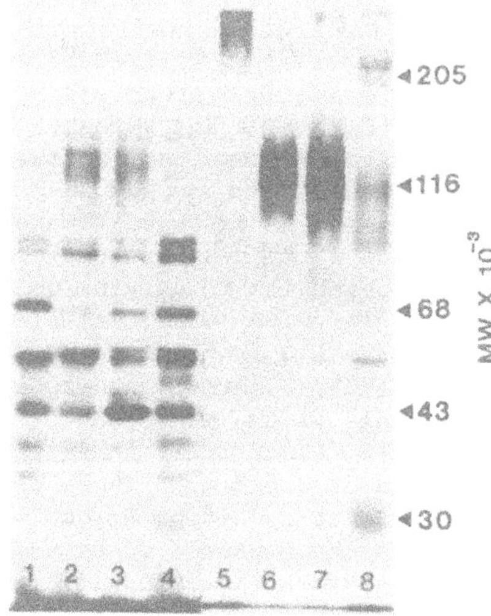

Figure 4. Fluorographs of one-dimensional gels (7.5% SDS-PAGE) of human oviductal culture medium labelled with L-[^{3}H] leucine (lanes 1-4) or L-[6-^{3}H] glucosamine (lanes 5-8). Tissue was obtained on day 3 of menses (lanes 1 and 5), day 8 of the cycle (lanes 2 and 6), midcycle (lanes 3 and 7) and day 21 of the cycle (lanes 4 and 8). Note the intensification of labelled secretory products during the follicular stage and midcycle.

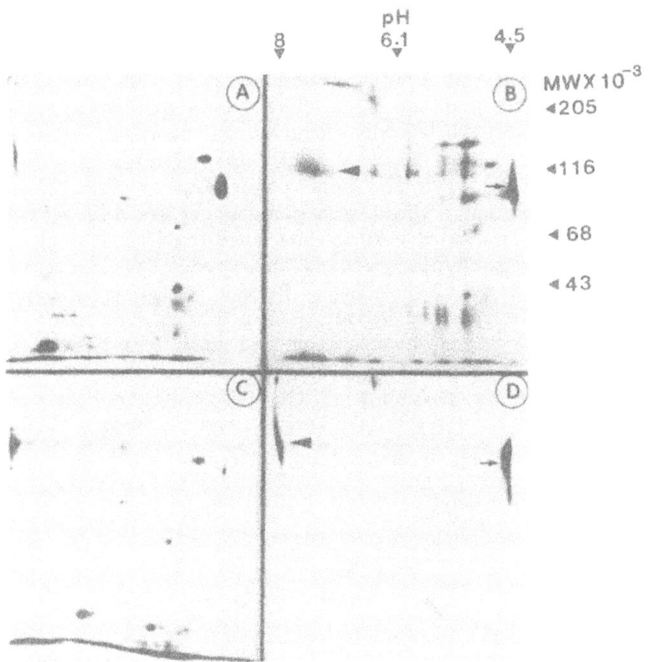

Figure 5. Fluorographs of two dimensional gels (7.5% acrylamide) of baboon ampulla culture media (Panels A and B) and human oviduct media (Panels C and D) labelled with L-[^{35}S] methionine (A and C) or L-[6-^{3}H] glucosamine (B and D). The major glycosylated proteins synthesized by both the baboon and human oviduct during the follicular stage or with estradiol treatment are the basic (arrowhead) and acidic (arrow) macromolecules that have similar electrophoretic mobilities in both species (Verhage and Fazleabas, 1988; Verhage *et al.*, 1988, reprinted with permission).

probably mediated through the steroid receptor systems (Verhage and Jaffe, 1986). Apparently, progesterone, acting through its own receptor system, decreases the nuclear binding of the estradiol-receptor complex, thus antagonizing the actions of estradiol without inducing the synthesis of major new macromolecules in the primate oviduct. This action of progesterone on the primate oviduct is in direct contrast to its action in the primate uterus where progesterone induces a typical progestational response (Brenner and Maslar, 1988) which results in the synthesis of progesterone-dependent macromolecules (Bell *et al.*, 1985; Fazleabas and Verhage, 1987a; Fazleabas *et al.*, 1988).

A concept that appears to be emerging is that the mammalian oviduct synthesizes and secretes major, estrogen-dependent macromolecules (Fazleabas and Verhage, 1986; Verhage and Fazleabas, 1988; Verhage *et al.*, 1988; Oliphant and Ross, 1982; Sutton *et al.*, 1986; Kapur and Johnson 1985; 1986). The precise role that these proteins might play in regulating early reproductive events is unknown; however, recent studies do suggest that they are important in early embryogenesis. Oviductal proteins become associated with the zona pellucida or fertilized ovum in the mouse (Kapur and Johnson, 1986) and hamster (Leveille *et al.*, 1987). The sulfated glycoproteins produced by the secretory oviductal epithelial cells of the rabbit may also play a role in regulating the maternal humoral system (Oliphant *et al.*, 1984). Recent studies by Gandolfi and Moor (1987) suggest that oviduct epithelial cells may produce specific factors that aid in embryonic development and maintain gamete viability.

It seems reasonable to assume that the major glycoproteins we have identified in the primate oviduct (Fazleabas and Verhage, 1986; Verhage and Fazleabas, 1988; Verhage *et al.*, 1988) may potentially be helpful in

Figure 6. One micron plastic section obtained from the region lining the lumen of baboon ampulla immunoreacted with the polyclonal antibody to the estradiol-induced baboon oviduct glycoproteins. Immunoperoxidase staining is limited to the discrete granular structure in the apical blebs. Staining is absent in ciliated cells. Magnification x540.

developing culture media more ideally suited toward supporting embryo development for *in vitro* fertilization programs and in research programs geared towards genetic manipulation of embryos and cryopreservation. However, additional information relating to the biochemical nature of these macromolecules is necessary. Towards this end, we have recently developed polyclonal antibodies to some of the antigens in the M_r 100,000-130,000 range of baboon estrogen-induced oviductal proteins. Our antibody preparation only immunoreacts with antigens present in oviductal flushings obtained from follicular stage or estradiol-treated ovariectomized baboons and is capable of immunoprecipitating labelled proteins from explant culture media of tissues from these two groups of animals (Verhage *et al.*, 1989). Immunocytochemical procedures demonstrate that the secretory epithelial cells of the estrogen-dominated baboon oviduct synthesizes these glycoproteins and stores them in apical secretory granules (Verhage *et al.*, 1989; Figure 6). Preliminary studies indicate that the electrophoretically similar glycoproteins secreted by the human oviduct at mid-cycle (Verhage *et al.*, 1988) also crossreact with the baboon polyclonal antibody directed towards this group of antigens (unpublished observations). The evidence that estrogen-dependent macromolecules of the mouse and hamster oviduct become associated with the fertilized ovum during transit (Kapur and Johnson, 1986; Leveille *et al.*, 1987), coupled with our studies in the primate, suggest that the proteins of the mammalian oviduct play a significant paracrine role in maintaining gamete viability.

UTERUS

The uterus, at the proper stage of the reproductive cycle, provides a hospitable and privileged environment for the early development of the blastocyst and the development of the conceptus following attachment and/or implantation. Associated with this fetal/maternal interaction are marked morphological and secretory changes in the endometrium that are modulated by the ovarian steroids. The secretions of the uterine endometrium may potentially play a significant role in initiating and maintaining a pregnancy (Bell, 1986, 1988) by acting in a paracrine manner (Healy and Hodgen 1983; Bell and Smith, 1988).

Unfortunately, it is not possible to study the role of specific primate endometrial proteins and their potential paracrine function during the very early stages of pregnancy, that is implantation in the human. However, based on the less than optimal pregnancy rates in women undergoing infertility therapy (Speroff *et al.*, 1979; Jones, 1983), an understanding of maternal/fetal interaction during implantation and the establishment of pregnancy is critical if the success of the present infertility therapies are to be improved. Therefore, we have chosen to pursue studies towards understanding maternal/fetal interaction in the baboon, a phylogenetically related non-human primate, that in many ways is the ideal primate model.

Morphology

The morphological changes of the baboon endometrium during the menstrual cycle and during steroid treatment of ovariectomized animals is very similar to that observed in the human (Verma, 1983; Cornillie *et al.*, 1985; Brenner and Maslar, 1988). During the follicular stage of the cycle or following 14 days of estradiol treatment, the endometrium is thick and endometrial glands are well developed. In contrast to the ovariectomized animal, the stroma is highly cellular and less compact and

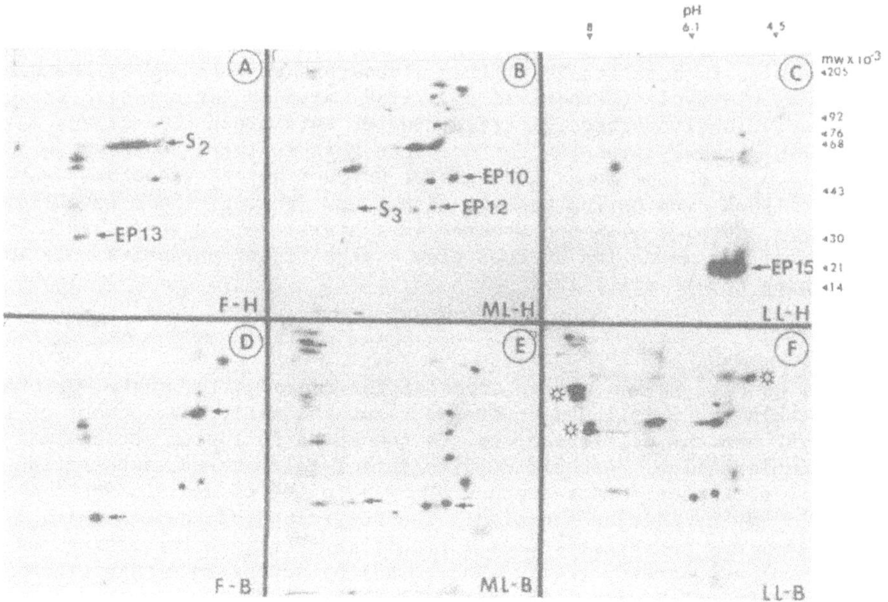

Figure 7. Fluorographs of two-dimensional gels (10% acrylamide) comparing endometrial culture medium labelled with L-[^{35}S] methionine from humans and baboons at the follicular (Panels A and D), mid-luteal (Panels b and e) and late luteal (Panels C and F) stages of the menstrual cycle. Note the similarity in the pattern of protein synthesis during the follicular and mid-luteal stages of the cycle and the marked difference during the luteal stage of the cycle. The human proteins are designated using previously published nomenclature (Strinden and Shapiro, 1983; Bell *et al.*, 1985) and electrophoretically similar proteins in the baboon are indicated by arrows.

both the glandular and surface epithelial cells increase in height. The apical tips of epithelial cells are characterized by the presence of numerous secretory granules. At mid-cycle or following 7 or 14 days of estrogen and progesterone treatment, the endometrium is very thick and the neck glands are long and prominent. The stroma is less compact compared to that observed during estrogen dominance, except around the prominent spiral arteries. The epithelial cells are columnar with basally located nuclei, and the cells are characterized by the presence of apical electron translucent granules filled with flocculent material. Large deposits of glycogen are present in the basal region. During the late luteal stage of the menstrual cycle or following 14 days of progesterone treatment to an estrogen-primed animal the endometrium is thinner than that observed in the estrogen-primed, estrogen plus progesterone treated animals. The stroma is more compact and the endometrial glands highly dilated. The epithelial cells appear to be shorter than that observed at mid-cycle; however, they are still columnar and contain basal nuclei. Apical granules are less conspicuous and glycogen deposition is still observed in both the apical and basal cytoplasm. The amount of glycogen present during the late luteal stage of the menstrual cycle is markedly diminished when compared with the progesterone treated endometrium. Accumulation of glycogen in the endometrium is a progesterone dependent process (Jaffe *et al.*, 1985), and the presence of glycogen in the steroid treated animal compared to the cycling animal may reflect continued steroid stimulation with out the normal decrease that is observed in the cycling animal prior to menses.

Secretory Proteins of the Baboon Endometrium

Studies on the *in vitro* secretory activity of the baboon uterine endometrium correlate with the observed morphological changes and further demonstrate that the general pattern of protein biosynthesis in this phylogenetically related non-human primate (Fazleabas and Verhage, 1987a) are similar to the changes described in the human endometrium (Strinden and Shapiro, 1983; Bell *et al.*, 1985, 1986a,b; Heffner *et al.*, 1986; Figure 7). Like the human (Bell *et al.*, 1989), the general secretory pattern of the baboon endometrium can be divided into two groups; a group (I) of proteins that are present throughout the menstrual cycle and show only minor cyclic variations in intensity, and a group (II) of proteins whose secretion is hormonally modulated and stage specific (Fazleabas and Verhage, 1987a; Fazleabas *et al.*, 1988). Group I proteins include several high molecular weight glycoproteins (M_r >100,000) and at least five additional lower molecular weight proteins (M_r 80,000 to 37,000; pI's 5.1 to 6.0; labelled 1 to 5 in Figure 8). Of these proteins 1 (M_r 76,000), 2 (M_r 66,000), 4 (M_r 40,000) and 5 (M_r 37,000) are synthesized and released by endometrial tissues obtained from untreated, long-term ovariectomized animals (Fazleabas *et al.*, 1988), thus suggesting that they do not require any hormonal stimulus for synthesis at baseline levels. However, in response to steroids the synthesis of these proteins are increased, suggesting that steroids do enhance their production (Fazleabas and Verhage, 1987a; Fazleabas *et al.*, 1988).

Group II proteins include those that are observed at specific stages of the menstrual cycle (Fazleabas and Verhage, 1987a) and can be induced by estrogen or with estrogen priming followed by treatment with progesterone (Fazleabas *et al.*, 1988). A baboon follicular stage protein (M_r 33,000, pI 7.6) is only observed in those ovariectomized baboons treated with estrogen for 7 or 14 days. Thus, this protein, which has similar electrophoretic properties to one synthesized by the human endometrium during the proliferative stage of the cycle (Bell *et al.*, 1986a), appears to be an estrogen-dependent protein (Figures 7 and 8).

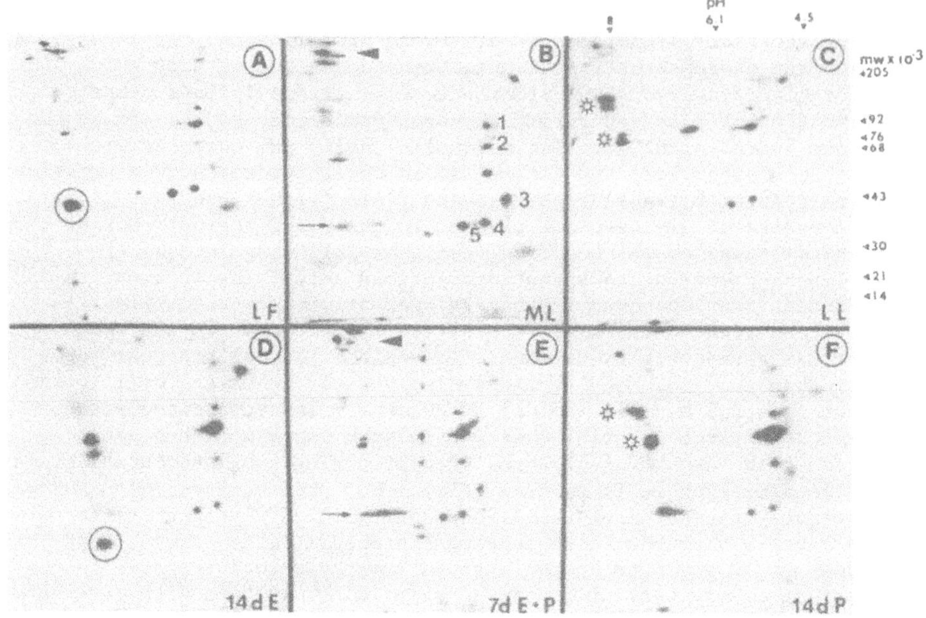

Figure 8. Fluorographs of two-dimensional gels (10% acrylamide) comparing endometrial culture medium labelled with L-[^{35}S] methionine from cycling baboons at the late follicular (A), mid-luteal (B), and late luteal (C) stages of the menstrual cycle with ovariatomized animals treated for 14 days with estradiol (D), 14 days of estradiol followed by 7 days of estradiol plus progesterone (F). Note the remarkable similarity in the pattern of protein biosynthesis between endometrium obtained from cycling animals compared with their corresponding steroid-treated counterparts (Fazleabas *et al.*, 1988, reprinted with permission).

A group of high molecular weight basic proteins (M_r > 200,000; arrowhead, Figure 7) is only prominent in the estrogen-primed, estrogen plus progesterone-treated animals and absent in the estrogen-primed animals treated with progesterone only. The synthesis of a single low molecular weight basic protein (M_r 40,000; arrow, Figure 8) is induced by progesterone regardless of the presence or absence of estrogen. This basic protein is electrophoretically similar to a previously described P-induced protein (S_3) isolated from human endometrium (Strinden and Shapiro, 1983).

Although the biological significance of primate endometrial proteins has yet to be elucidated, it is apparent from our studies in the baboon (Fazleabas and Verhage, 1987a; Fazleabas *et al.*, 1988) and the human (Bell *et al.*, 1985, 1986a,b) that the pattern of protein synthesis by the primate endometrium is the most complex during the early to mid-luteal stages of the menstrual cycle. Presumably, these proteins play an important regulatory role in early events associated with embryonic development and implantation. Endometrial secretory activity during the preimplantation period also coincides with high serum levels of both estrogen and progesterone. The role that elevated levels of estrogen may play during the luteal stage in preparing the endometrium for implantation is as yet unclear. However, our results clearly demonstrate that the synthetic profile is different in animals first primed with estrogen and then treated with estrogen and progesterone compared to those treated with progesterone alone (Fazleabas *et al.*, 1988). This change in synthetic pattern may be due to the loss of the synergistic action of estrogen and progesterone or perhaps the failure to maintain the progesterone receptor system in the absence of estrogen. The synergistic action of estrogen and progesterone on the modulation of endometrial proteins in the progestational uterus has been well documented in the pig (Basha *et al.*, 1980; Fazleabas *et al.*, 1982). In the primate, progesterone is the principal mediator of secretory endometrial development, and estrogen appears to be essential for inducing progesterone receptors (Bayard *et al.*, 1978; Levy *et al.*, 1980). Antagonism of estrogen during the luteal stage in the human lowers progesterone receptor concentrations and significantly affects secretory endometrial development (Fritz *et al.*, 1987). Although the overall endometrial synthetic activity is altered in the absence of estrogen, in the baboon, progesterone does sustain the synthesis of three basic proteins (Mr 88,000, 66,000, and 40,000), thus suggesting that although progesterone receptor concentrations may be diminished, progesterone function is still maintained. Our studies on endometrial protein biosynthesis in the baboon suggest that estrogen and progesterone act synergistically to sustain maximal secretory activity and further substantiate the previous morphological observations in the non-human primate (Brenner *et al.*, 1983) and humans (Fritz *et al.*, 1987).

In summary, these studies on the baboon uterine endometrium demonstrate that the morphological changes associated with the menstrual cycle correlate with *in vitro* secretory changes of the tissue. Furthermore, sequential treatment of ovariectomized baboons resulting in physiological steroid levels essentially duplicate morphological and secretory changes observed during the normal cycle. The ability to regulate the baboon endometrium with exogenous hormone treatments that result in similar changes observed during the menstrual cycle have now provided the means to purify proteins of interest and attempt to elucidate their potential biological functions.

COMPARATIVE STUDIES ON THE BABOON AND HUMAN UTERUS

General

Baboon group I proteins 2 (M_r 66,000, 3 (M_r 46,000) and 5 (M_r 37,000) are electrophoretically similar to human endometrial proteins EP8, EP10, and EP12 described by Bell *et al.* (1985). In addition, a protein similar in molecular weight to one first described by Strinden and Shapiro (1983) as a progesterone-induced protein, S_3, is also observed as a broad streak at the basic end of the fluorograph from the late luteal stage of the baboon (Figure 7). Baboon protein 2 also has electrophoretic characteristics similar to a protein found in human uterine fluid (Sylvan *et al.*, 1981; MacLaughlin and Richardson, 1983). This protein was the only readily identifiable nonserum product observed at both stages of the menstrual cycle and appeared to be enhanced during the secretory phase. Our *in vitro* studies confirm this finding (Fazleabas and Verhage, 1987a; Fazleabas *et al.*, 1988); in addition, analysis of baboon uterine flushings indicates that this is the only product of nonserum origin that can be identified on silver-stained polyacrylamide gels of baboon uterine flushings (Fazleabas and Verhage, 1987b), Although the general patterns of protein biosynthesis throughout the cycle and the secretion of a specific product during the follicular stage are similar in both the baboon and the human, the proteins released *in vitro* during the late luteal stage are remarkably different (Figure 7). The human endometrium, in culture, incorporates 40-50% of the radioactive label into a product of M_r 25,000-28,000 under reducing conditions. This protein has been termed EP15 or pregnancy-associated endometrial α2-globulin (α2-PEG) by Bell *et al.* (1985, 1986a,b). Further characterization of this molecule suggests that it is similar to endometrial protein PEP (Joshi *et al.*, 1980), since PEP cross-reacts immunologically with PP_{14} (Bell and Bohn, 1986; Julkunen *et al.*, 1986). This protein is not detected in the baboon. Instead, under progesterone dominance, the baboon endometrium predominantly produces a group of basic proteins (M_r 80,000-90,000, M_r 59,000, M_r 62000, and M_r 40,000) and an acidic protein (M_r 130,000).

α2-PEG/EP15 has been characterized as a polymorphic dimeric glycoprotein which has extensive sequence homology with the milk proteins, the β-lactoglobulins (Julkunen *et al.*, 1988; Bell *et al.*, 1989), and retinol (Bell *et al.*, 1989). Although structurally similar to these two molecules, its putative biological function appears to be immunosuppression (Bell *et al.*, 1989). In spite of being the major protein product of the human glandular epithelium during the late luteal stage of the cycle, no immunologically similar molecule is present in the baboon (Fazleabas and Verhage, 1987a; Figure 9). The absence of a similar protein in the baboon illustrates the problems associated with selecting putative analogs between species based on physiochemical criteria. It is possible that another electrophoretically and/or immunologically unrelated molecule may perform a similar function. The fact that synthesis of α_2-PEG/EP15 in the human is restricted primarily to early pregnancy suggests that this molecule mediates some function of the glandular epithelium during implantation. Based on the differences in the implantation process between the baboon and human (Hearn, 1986; Ramsey *et al.*, 1976), it is intriguing to speculate that this dramatic difference in endometrial secretory activity during implantation between these two species is reflective of this process and that proteins produced during this period may have different regulatory roles.

Placental protein 12 (PP_{12}) is biochemically and immunologically similar to EP14 or pregnancy-associated endometrial α_1-globulin (α1-PEG, Bell and Bohn, 1986). This molecule is an insulin-like growth factor binding protein (IGF-BP) which is similar to the low molecular weight (Mr 30,000-38,000) IGF-BP's isolated from human amniotic fluid (Drop *et al.*, 1984; Povoa *et al.*, 1984; Baxter *et al.*, 1987). This IGF-BP is found in the non-pregnant human endometrium (Wahlstrom and Seppala, 1984; Waites *et al.*, 1988a). However, its site of synthesis and/or storage appears to be the glandular epithelium of the late luteal stage endometrium when a polyclonal antibody to PP_{12} is employed for immunocytochemistry (Walstrom and Seppala, 1984) and predecidualized stromal cells when monoclonal antibodies are used in this procedure (Figure 10;Waites *et al.*, 1988a). We have also identified an immunologically similar IGF-BP in the non-pregnant baboon endometrium during the late luteal stage or with progesterone treatment following estrogen priming (Fazleabas *et al.*, 1989a; see next section for discussion). Although readily identifiable on Western blots of tissue culture media (Fazleabas and Verhage, 1987a; Fazleabas *et al.*, 1989a) and immunolocalized within the glandular epithelium during the luteal stage (Figures 10A and B; Fazleabas *et al.*, 1989a), this protein does not constitute a major radiolabelled secretory product on fluorographs of baboon endometrial culture media. In contrast, the late luteal human endometrium does secrete an identifiable, radiolabelled IGF-BP, albeit minor, during short term culture (Bell *et al.*, 1985). The synthesis of this protein can be maximized in culture following 3 to 4 days of progesterone exposure *in vitro* (Rutanen *et al.*, 1986). Although similar long term organ culture studies have not been carried out with baboon endometrial tissue, the localization of IGF-BP in this species implies a potentially important role for this molecule

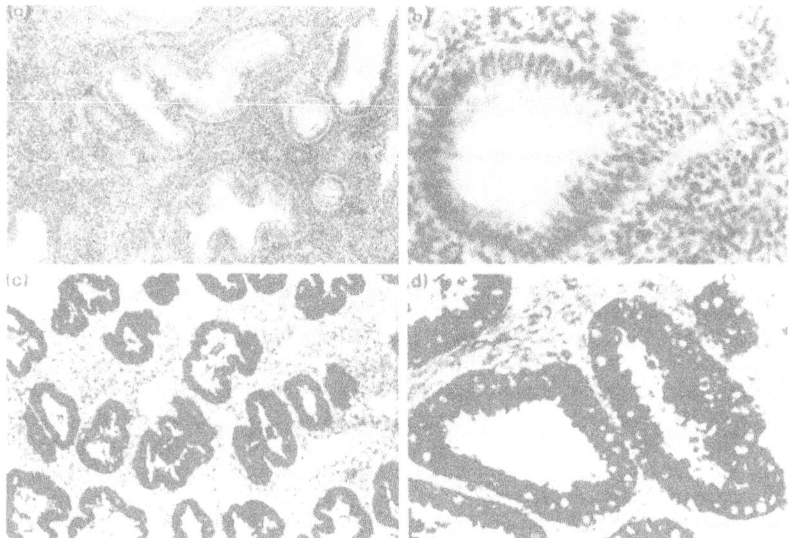

Figure 9. Immunocytochemical localization of α_2-PEG/EP15 in late luteal endometrial tissue from baboons and humans using the monoclonal antibody C6H11. Note that the baboon endometrium shows no immunoactivity (Panels A and B; Magnification x92 and x363), while the glandular epithelium of the human endometrium stains intensely (Panels C and D; Magnification x82 and x328).

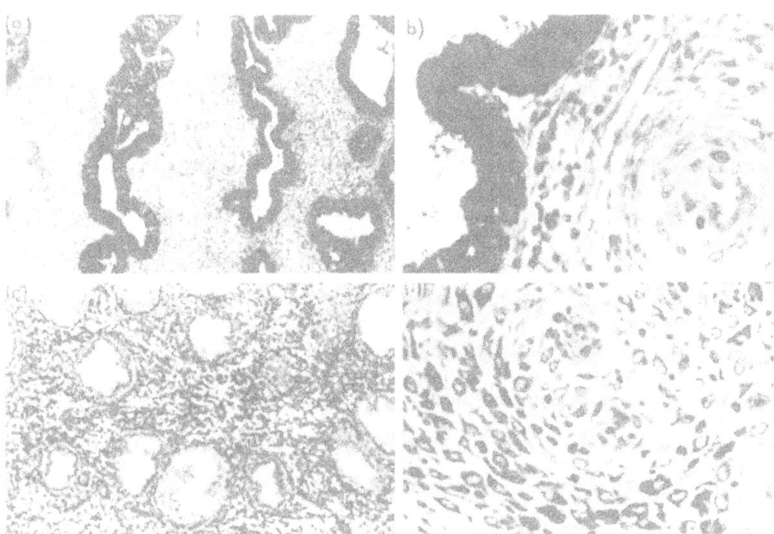

Figure 10. Immunocytochemical localization of α_1 PEG/IGF-BP in late luteal endometrial tissue from baboons and humans using monoclonal antibody B2H10. Note that the epithelium of the deep glands of the baboon endometrium stain intensely (Panels A and B; Magnification x92 and x363) while in the human, stromal staining predominates at the same stage of the cycle (Panels C and D; Magnification x82 and x328.

during early pregnancy. The absence of a readily identifiable secretory protein in whole tissue culture experiments illustrates the potential limitations of this technique, whereby the limited synthesis of a particular product due to regional restriction or association with a limited cellular population could easily be overlooked. IGF-BP is an excellent example of this particular problem, since this protein, in association with IGF, has the potential of having a paracrine role in either stimulating or inhibiting IGF-action (Elgin *et al.*, 1987; Busby *et al.*, 1988). The localized action and limited production of this protein in the baboon endometrium may indeed reflect a specific regulatory role for this protein during implantation and early pregnancy (see below).

Prolactin, another quantitatively minor product of the endometrium, has been detected as an immunoreactive protein in endometrial explant culture media from the luteal stage baboon (Fazleabas and Verhage, 1987a), and its rate of secretion in long term progesterone-treated cultures has been correlated with predicidualization in the human (Maslar *et al.*, 1986). Incorporation studies have failed to detect labelled prolactin in either the human or baboon; however, Heffner *et al.*, (1986) have reported the synthesis of a glycosylated form of prolactin by the human endometrium during the late luteal stage. Decidual prolactin by the human endometrium during the late luteal stage. Decidual prolactin represents the major extrapituitary source of prolactin during pregnancy in the human (Riddick and Daly, 1982; Healy 1984; Riddick 1985) and non-human primate (Maslar *et al.*, 1988), and certain lines of evidence support the concept that prolactin in pregnancy may also act as a paracrine hormone (Tyson, 1982; Healy, 1984; Riddick, 1985).

In addition to IGF-BP (PP_{12}) and prolactin, which are minor synthetic products during the late luteal stage and become the major secretory products of the baboon and human decidua (Bell and Smith, 1988; Fazleabas *et al.*, 1989b), the baboon endometrium also produces three other proteins in culture that are initially isolated from human term placenta and called placental proteins (PP). Subsequent studies have demonstrated that these proteins are synthesized by the endometrium or both the endometrium and the oviduct (Fazleabas and Verhage, 1987; Verhage *et al.*, 1988). PP_{16} is secreted under the influence of progesterone. In contrast, PP_4, which has cDNA sequence homology with the lipocortin gene family (Grundman *et al.*, 1988; also see Chapter 1), and PP_7, which has been identified as the enzyme glutathione S-transferase (Bohn, 1985), are secreted throughout the menstrual cycle. Furthermore, PP_4, which also has similar electrophoretic mobility to the human endometrial protein EP_{12} (Bell *et al.*, 1985), and PP_{16} can also be identified in fluorographs of baboon endometrial culture media (Fazleabas and Verhage, 1987a). The functional significance associated with the presence of so-called "placental proteins" in both human and baboon endometrial culture media is intriguing, and the possibility arises that some of the trophoblastic membrane proteins originate in the endometrium and probably play an essential role during implantation and early embryonic development.

IGF-BP Synthesis in the Non-Pregnant and Pregnant Baboon

The somatomedins, IGF-I and IGF-II, are thought to act via endocrine, paracrine and/or autocrine mechanisms (Underwood and D'Ercole, 1984; Underwood *et al.*, 1986) and their binding proteins are thought to control their action (Hintz, 1984). Recent evidence (Elgin *et al.*, 1987; Busby *et al.*, 1988) suggests that human amniotic fluid contains two forms of IGF-BP's which have virtually identical physiochemical properties but distinctly different biological properties. Although no structural differences were identified, the enhancement and inhibition of DNA synthesis in porcine smooth muscle cells by two proteins isolated following ion-exchange chromatography of human amniotic fluid was associated with the difference in membrane adherence of these two forms of binding protein (Busby *et al.*, 1988).

The close association between the developing trophoblast and decidualized endometrium together with evidence for somatomedin production by placental tissue (Fant *et al.*, 1986) suggest that decidual IGF-BP in the baboon and human may regulate feto-placental growth. This protein is a minor synthetic product of both the baboon and human endometrium during the luteal stage of the menstrual cycle (Fazleabas and Verhage, 1987a; Bell *et al.*, 1986a). However, during pregnancy IGF-BP becomes the major secretory protein of the stromal cells of the decidua in both the human (Bell *et al.*, 1988; Waites *et al.*, 1988b) and the baboon (Fazleabas *et al.*, 1989b). Studies in the baboon demonstrate that α_1-PEG/IGF-BP is immunologically and biochemically identical to the human protein (Fazleabas *et al.*, 1989b), and that IGF-BP is quantitatively the major synthetic product of third trimester baboon decidua. Biochemical characterization of the baboon decidual IGF-BP indicate that it is an acidic polypeptide (M_r 33,000) consisting of two isoelectric forms of identical molecular weight. This molecule binds IGF-I with a Kd of 1.14-1.83 nM and forms a complex with ^{125}I-labelled IGF-I which has a Mr of 36,000 under reducing conditions (Fazleabas *et al.*, 1989b). The cDNA coding for this protein hybridizes a single message transcript of 1.6 kb on Northern blots of total RNA isolated from decidual tissue and late luteal baboon and human endometrium (Fazleabas *et al.*, 1989a).

The IGF-BP is immunolocalized in the perinuclear region of the hypertrophied stromal cells of the baboon decidua (Figure 11C and D; Fazleabas *et al.*, 1989b), similar to that observed in the decidua of human pregnancy (Figures 11E and F; Waites *et al.*, 1988b). However, during the menstrual cycle the immunolocalization of the protein is different between the baboon (Fazleabas *et al.*, 1989a) and the human (Waites *et al.*, 1988a; Figure 10). The major site of synthesis of IGF-BP in the non-pregnant baboon is the glandular epithelium, whereas in the human, epithelial staining is minimal and the predicidualized stromal cell shows the most immunoreactivity. Thus, the principal difference between these two species is that, in the human, the major site of localization is stromal, whereas in the baboon, the synthesis switches from glandular to stromal with pregnancy. The functional significance of this is unknown. However, it has been suggested that IGF-BP may play a role in either regulating trophoblast growth in the human or alternatively permit proliferating stromal fibroblasts to respond to endogenous IGF (Bell and Smith, 1988; Elgin *et al.*, 1987). The more recent evidence from Clemmons group (Busby *et al.*, 1988) indicating that two different biological forms of IGF-BP exist in human amniotic fluid and that their net effect is to regulate cellular response to IGF-I raises the intriguing possibility that, in the baboon, the switch in synthesis from glandular to stromal may represent the synthesis of the two different forms at specific stages of the cycle and that this may potentially play a role in implantation in this species.

The development of the feto-placental unit in humans is associated with dramatic alterations in endometrial differentiation during early pregnancy (Bell, 1988). It has been suggested that the interactions between the developing embryo and its membranes with the endometrium is mediated via endometrial secretory products (Bell, 1986; Bell, 1988) acting in a paracrine manner (Bell and Smith, 1988; Healy and Hodgen, 1983). However, due to moral and ethical limitations, the nature by

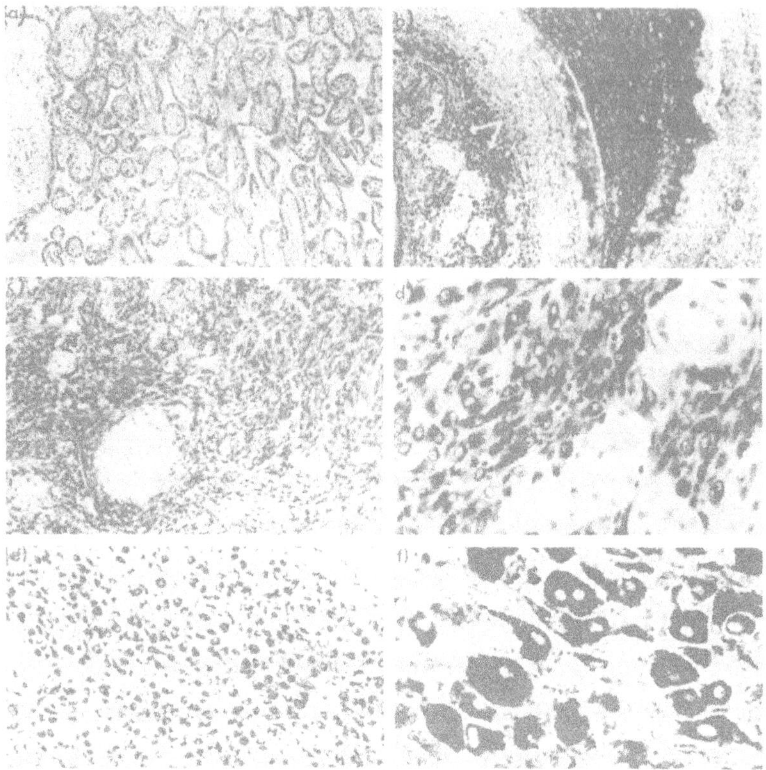

Figure 11. Photomicrographs of baboon placental villi (Panel A; Magnification x54), chorio-amniotic membranes with adhering decidua (Panel B; Magnification x70), decidua (Panels C and D; Magnification x70 and x275) and human decidua (Panels E and F; Magnification x82 and x328) obtained during the third trimester (baboon) and first trimester (human). The sections were immunostained with monoclonal antibodies to α_1-PEG/IGF-BP. Note that the decidua or hypertrophied stromal cells in both the baboon (arrowed; Panels B and D) and human (Panels E and F) stain intensely. This is in contrast to the late luteal stage of the menstrual cycle, where the immunolocalization of IGF-BP is distinctly different in these two species.

which endometrial secretory products interact with the developing conceptus during the initial stages of implantation cannot be studied in the human. Since we have demonstrated that endometrial protein biosynthesis in the baboon (Fazleabas and Verhage, 1987a; Fazleabas *et al.*, 1988) is very similar to the human, except in the late luteal stage, and since the major decidual secretory product in both human and baboon decidua is an IGF-BP derived from an identical cellular source, the baboon is a reliable model to study maternal-fetal interrelationships in early pregnancy. Although the baboon and human do demonstrate morphological differences in implantation (Ramsey *et al.*, 1976), it is tempting to speculate that the difference in the site of synthesis of IGF-BP during the luteal stage of the menstrual cycle may play a role in controlling trophoblast invasion. During implantation in the human, trophoblast cells rapidly penetrate subepithelial stroma and the spiral arteries, the major sites of synthesis of IGF-BP (Waites *et al.*, 1988b). Since implantation is less invasive in the baboon (Hearn, 1986), one might speculate that trophoblast penetration needs to be more stringently controlled in this species. If IGF-BP does play a role in this regulation, perhaps the optimal sites of synthesis for this protein would be the glandular epithelium of the deep functionalis and basalis regions.

In summary, comparative studies from our laboratories have clearly demonstrated that the baboon and human endometrium synthesize and secrete a number of electrophoretically and immunologically similar proteins, principal amongst which is an IGF-BP. Hypotheses concerning the function of this protein during the early post-implantation stages in the human can only be deduced to a limited extent. Even though differences exist in the type of implantation in the baboon as compared to the human, the fact that a histologically defined cell type quantitatively secretes one major common product, implies that questions concerning fetal-maternal regulations immediately post-implantation in the primate can now be addressed. Furthermore, the intriguing observation that the immunocytochemical localization of this protein, using the same monoclonal antibody, is different in the non-pregnant human compared to the non-pregnant baboon, may provide insight into the role of the conceptus in altering the cellular site of gene expression during implantation and pregnancy.

ACKNOWLEDGEMENTS

These studies were supported by National Institutes of Health grants HD 21991 (ATF), HD 20571 (HGV) and funds from the Medical Research Council and Wellcome Trust (SCB). The authors also wish to acknowledge the contributions of Dr. Gillian Waites, Dr. Melinda Boice, Ms. Kathy Donnelly and Ms. Patty Movrogianis, and thank Ms. Margarita Guerrero for typing this manuscript.

REFERENCES

Basha S.M.M., Bazer F.W., Giesert R.D. and Roberts R.M. 1980. Progesterone-induced uterine secretions in pigs. Recovery from pseudopregnant and unilaterally pregnant gilts. J. Anim. Sci. 50:113.

Baxter R.C., Martin J.L. and Wood M.H. 1987. Two immunoactive binding proteins for insulin-like growth factors in human amniotic fluid: relationship to fetal maturity. J. Clin. Endocrinol. Metab. 65:423.

Bayard F., Damilano S., Robel P. and Baulieu E.E. 1978. Cytoplasmic and nuclear estradiol and progesterone receptors in human endometrium. J. Clin. Endocrinol. Metab. 46:635.

Bell S.C. 1986. Secretory endometrial and decidual proteins: studies and clinical significance of a maternally derived group of pregnancy-associated serum proteins. Human Reprod. 1:129.

Bell S.C. 1988. Secretory endometrial/decidual proteins and their function in early pregnancy. J. Reprod. Fertil. Supp. 36:109.

Bell S.C. and Bohn H. 1986. Immunochemical and biochemical relationship between human pregnancy-associated secreted endometrial $\alpha 1$- and $\alpha 2$-globulins ($\alpha 1$- and $\alpha 2$- PEG) and soluble placental proteins 12 and 14 (PP_{12} and PP_{14}). Placenta 7:283.

Bell S.C., Fazleabas A.T. and Verhage H.G. 1989. Comparative aspects of secretory proteins of the endometrium and decidua in human and non-human primates - primate models for the study of the function of the endometrium and decidua, "Serono Symposium on Blastocyst Implantation" Plenum Press, N.Y. (in press).

Bell S.C., Hales M.W., Patel S.R., Kirwan P.H. and Drife J.O. 1985. Protein synthesis and secretion by the human endometrium and decidua during early pregnancy. Br. J. Obst. Gynecol. 92:793.

Bell S.C., Patel S., Kirwan P.H. and Drife J.O. 1986a. Protein synthesis and secretion by the human endometrium during the menstrual cycle and effect of progesterone *in vitro*. J. Reprod. Fertil. 77:221.

Bell S.C., Patel S., Hales M.W., Kirwan P.H. and Drife J.O. 1986b. Immunochemical detection and characterization of pregnancy associated endometrial $\alpha 1$- and $\alpha 2$- globulins secreted by human endometrium and decidua. J. Reprod. Fertil. 74:261.

Bell S.C. and Smith S. 1988. The endometrium as a paracrine organ. In: "Contemporary Obstetrics and Gynecology" (G.V.P. Chamberlain, ed.), Butterworths Scientific Ltd., London. p.273.

Bohn H. 1985. Biochemistry of placental proteins. In: "Proteins of the Placenta. 5th Int. Congress on Placental Proteins", (P. Bischof and A. Klopper, eds.), Karger, Basel, p.1.

Brenner R.M., Carlise K.S., Hess D.L., Sandow B.A. and West N.B. 1983. Morphology of the oviducts and endometria of Cynomolgus macaques during the menstrual cycle. Biol. Reprod. 29:1989.

Brenner R.M. and Maslar I.A. 1988. The primate oviduct and endometrium. In: "The Physiology of Reproduction, Vol. 1", (E. Knobil and J.D. Neill, eds.), Raven Press, N.Y., p. 303.

Brenner R.M., West N.B., Norman R.L., Sandow B.A. and Verhage H.G. 1979. Progesterone suppression of the estradiol receptor in the reproductive tract of macaques, cats and hamsters. In: "Steroid Hormone Receptors Systems", (W.W. Leavitt and J.H. Clark, eds.), Plenum Press, N.Y., p.173.

Busby W.H., Jr., Klapper D.G. and Clemmons D.R. 1988. Purification of a 31,000-dalton insulin-like growth factor binding protein from human amniotic fluid. J. Biol. Chem. 263:14203.

Cornillie F.J., Lauweryns J.M. and Brosens I.A. 1985. Normal human endometrium; an ultrastructural survey. Gynecol. Obstet. Invest. 20:113.

Drop S.L.S., Kortleve D.J. and Guyda H.J. 1984. Isolation of a somatomedin -binding protein from pre-term amniotic fluid. Development of a radioimmunoassay. J. Clin. Endocrinol. Metab. 59:899.

Elgin R.G., Busby W.H., Jr. and Clemmons D.R. 1987. An insulin-like growth factor (IGF) binding protein enhances the biologic response to IGF. Proc. Natl. Acad. Sci. (USA) 84:3254.

Fant M., Munro H. and Moses A. 1986. An autocrine/paracrine role for insulin-like growth factors in the regulation of human placental growth. J. Clin. Endocrinol. Metab. 63:499.

Fazleabas A.T., Bazer F.W. and Roberts R.M. 1982. Purification and properties of a progesterone-induced plasmin/trypsin inhibitor from uterine secretions of pigs and its immunocytochemical localization in the pregnant uterus. J. Biol. Chem. 257:6886.

Fazleabas A.T., Jaffe R.C., Verhage H.G., Waites G.T. and Bell S.C. 1989a. An insulin-like growth factor binding protein (IGF-BP) in the baboon (*Papio anubis*) endometrium: synthesis, immunocytochemical localization and hormonal regulation. Endocrinology 124:2321.

Fazleabas A.T., Miller J.B. and Verhage H.G. 1988. Ssynthesis and release of estrogen- and progesterone-dependent proteins by the baboon (*Papio anubis*) uterine endometrium. Biol. Reprod. 39:729.

Fazleabas A.T. and Verhage H.G. 1986. The detection of oviduct-specific protein in the baboon (*Papio anubis*). Biol. Reprod. 35:455.

Fazleabas A.T. and Verhage H.G. 1987a. Synthesis and release of polypeptides by the baboon (*Papio anubis*) uterine endometrium in culture. Biol Reprod. 37:979.

Fazleabas A.T. and Verhage H.G. 1987b. A simple technique for sampling the uterine cavity of the baboon. Theriogenology 27: 645.

Fazleabas A.T., Verhage H.G., Waites G.T. and Bell S.C. 1989b. Characterization of an insulin-like growth factor binding protein (IGF-BP), analogous to human pregnancy-associated secreted endometrical α_1-globulin (α_1-PEG), in decidua of baboon (*Papio anubis*) placenta. Biol Reprod. 40:73.

Fritz M.A., Westfahl P.K. and Graham R.L. 1987. The effect f luteal phase estrogen antagonism on endometrial development and luteal function in women. J. Clin. Endocrinol. Metab. 65:1006.

Gandolfi F. and Moor R.M. 1987. Stimulation of early embryonic development in the sheep by co-culture with oviduct epithelial cells. J. Reprod. Fertil. 81:23.

Grundmann U., Abel K-J., Bohn H., Lobermann H., Lottspeich F. and Kupper H. 1988. Characterization of cDNA encoding human placental anti-coagulant protein (PP_4): homology with the lipocortin family. Proc. Natl. Acad. Sci. (USA) 85:3708.

Healy D.L. 1984. The clinical significance of endometrial prolactin. Aust. and N.Z. J. Obstet. Gynecol. 24:111.

Healy D.L. and Hodgen G.D. 1983. The endocrinology of human endometrium. Obstet. Gynecol. Survey 38:509.

Hearn J.P. 1986. The embryo-maternal dialogue during early pregnancy in primates. J. Reprod. Fertil. 76:809.

Heffner L.J., Iddenden D.A., and Lyttle R.C. 1986. Electrophoretic analyses of secreted human endometrial proteins: identification and characterization of luteal phase prolactin. J. Clin. Endocrinol. Metab. 62:1288.

Hintz R.L. 1984. Plasma forms of somatomedin and the binding protein phenomena. Clinics Endocrinol. Metab. 13:31.

Jaffe R.C., Stevens D.M. and Verhage H.G. 1985. The effects of estrogen and progesterone on glycogen and the enzymes involved in its metabolism in the cat uterus. Steroids 45:453.

Jones H.W. Jr. 1983. Factors influencing implantation and maintenance of pregnancy following embryo transfer. In: "Fertilization of the Human Egg *In Vitro*", (H.M. Beier and H.R. Linder, eds.), Springer-Verlag, Berlin. p.293.

Joshi S.G., Ebert K.M. and Swartz D.P. 1980. Detection and synthesis of a progesterone-dependent protein in the human endometrium. J. Reprod. Fertil. 59:273.

Julkunen M., Raikar R.S., Joshi S.G., Bohn H., and Seppala M. 1986. Placental protein 14 and progesterone-dependent endometrial protein are immunologically indistinguishable. Human Reprod. 1:7.

Julkunen M., Seppala M. and Janne O.A. 1988. Complete amino acid sequence of human placental protein 14: a progesterone-regulated uterine protein homologous to β-lactoglobulins. Proc. Natl. Acad. Sci. (USA) 85:8845.

Kapur R.P. and Johnson L.V. 1985. An oviductal fluid glycoprotein associated with ovulated mouse ova and early embryos. Develop Biol. 112:89.

Kapur R.P. and Johnson L.V. 1986. Selective sequestration of an oviductal fluid glycoprotein the perivitelline space of mouse oocytes and embryos. J. Exp. Zool. 238:249.

Leveille M-C., Roberts K.D., Chevalier S., Chapdelaine A. and Bleau G. 1987. Uptake of an oviductal antigen by the hamster zona pellucida. Biol. Reprod. 36:227.

Levy C., Robel P., Gautray J.P., De Brux J., Verma U., Descomps B., Baulieu E.E. and Eychenne B. 1980. Estradiol and progesterone receptors in human endometrium: normal and abnormal menstrual cycles and early pregnancy. Am. J. Obstet. Gynecol. 136:646.

MacLaughlin D.T. and Richardson G.S. 1983. Analysis of human uterine luminal fluid proteins following radiolabelling by reductive methylation: comparison of proliferative and secretory phase samples. Biol. Reprod. 29:783.

Maslar I.A. and Ansbacher R. 1986. Effects of progesterone on decidual prolactin production by organ cultures of human endometrium. Endocrinology 118:2101.

Maslar I.A., Lazur J.J., Normamn R.L. and Spies H.G. 1988. Amniotic fluid and decidual prolactin during pregnancy in rhesus macaques. Biol. Reprod. 38:1067.

Odor D.L., Gaddum-Rosse, P. and Rumery, R.E., 1983, Secretory cells of the oviduct of the pig tailed macaque, Macaca nemestrina, during the menstrual cycle and after estrogen treatment. Am. J. Anat. 1166:149.

Oliphant G. and Ross P.R. 1982. Demonstration of production and isolation of three sulfated glycoproteins from the rabbit oviduct. Biol Reprod. 26:537.

Oliphant G., Reynolds A.B., Smith P.F., Ross P.R. and Marta J.S. 1984. Immunochemical localization of hormone-induced synthesis of the sulfated oviductal glycoproteins. Biol. Reprod. 31:165.

Povoa G., Enberg G., Jornvall H. and Hall K. 1984. Isolation and characterization of a somatomedin-binding protein from mid-term human amniotic fluid. Eur. J. Biochem. 144:199.

Ramsey E.M., Houston M.L. and Harris J.W.S. 1976. Interactions of the trophoblast and maternal tissues in three closely related primate species. Am. J. Obstet. Gynecol. 124:647.

Riddick D.H. and Daly D.C. 1982. Decidual prolactin production in human gestation. Seminars in Perinatol. 6:229.

Riddick D.H. 1985. Regulation and physiological relevance of non-pituitary prolactin In: "Prolactin Basic and Clinical Correlates," (R.M. MacLeod, M.O. Thomes and V. Scapagrini, eds.), Liviana Press, Padova. p.463.

Rutanen E-M., Koistinen R., Sjoberg J., Julkunen M., Wahlstrom T., Bohn H. and Seppala M. 1986. Synthesis of placental protein 12 by the human endometrium. Endocrinology 118:1067.

Speroff L., Glass R.H. and Kase N.G. 1979. Induction of ovulation with clomiphene and human menopausal gonadotrophins. In: "Clinical Gynecological Endocrinology and Infertility," (L. Speroff, R.H. Glass and N.G. Kase, eds.), The Williams and Wilkins Co., Baltimore. p.214.

Strinden S.T. and Shapiro S.S. 1983. Progesterone-altered secretory proteins from cultured human endometrium. Endocrinology 112:862.

Sutton R., Nancarrow C.D. and Wallace A.L.C. 1986. Estrogen and seasonal effects on the production of an estrus-associated glycoprotein in oviductal fluid of sheep. J. Reprod. Fertil. 77:645.

Sylvan P.E., MacLaughlin D.T., Richardson G.S., Scully R.E. and Nikrui N. 1981. Human uterine fluid proteins associated with secretory phase endometrium: progesterone-induced proteins? Biol. Reprod. 24:423.

Tyson J.E. 1982. The evolutionary role of prolactin in mammalian osmoregulation: effects on fetoplacental hydromineral transport. Seminars in Perinatol. 6:216.

Underwood L.E. and D'Ercole A.J. 1984. Insulin and insulin-like growth factors/ somatomedins in fetal and neonatal development. Clinics Endocrinol. Metab. 13:69.

Underwood L.E., D'Ercole A.J., Clemmons D.R. and Van Wyk J.J. 1986. Paracrine functions of somatomedins. Clinics Endocrinol. Metab. 15:59.

Verhage H.G., Bareither M.L., Jaffe R.C. and Akbar M. 1979. Cyclic changes in ciliation, secretion and cell height of the oviductal epithelium in women. Am. J. Anat. 156:505.

Verhage H.G., Boice M.L., Mavrogianis P., Donnelly K. and Fazleabas A.T. 1989. Immunological characterization and immunocytochemical localization of oviduct-specific glycoproteins in the baboon (*Papio anubis*). Endocrinology 124:2464.

Verhage H.G. and Fazleabas A.T. 1988. The *in vitro* synthesis of estrogen-dependent proteins by the baboon (*Papio anubis*) oviduct. Endocrinology 123: 552.

Verhage H.G., Fazleabas A.T. and Donnelly K. 1988. The *in vitro* synthesis and release of proteins by the human oviduct. Endocrinology 122:1639.

Verhage H.G. and Jaffe R.C. 1986. Hormonal control of the mammalian oviduct; morphological features and the steroid receptor systems. In: "The Fallopian Tube: Basic Studies and Clinical Contributions", (A.M. Siegler, ed.), Futura, N.Y. p.107.

Verma V. 1983. Ultrastructural changes in human endometrium at different phases of the menstrual cycle an their functional significance. Gynecol. Obstet. Invest. 15:193.

Wahlstrom T. and Seppala M. 1984. Placental protein 12 (PP_{12}) is induced in the endometrium by progesterone. Fertil. Steril. 41:781.

Waites G.T., James R.F.L. and Bell S.C. 1988a. Immunohistological localization of human endometrial secretory proteins "pregnancy-associated endometrial secretory α1-globulin (α1-PEG), an insulin-like growth factor binding protein, during the menstrual cycle. J. Clin. Endocrinol. Metab. 67:1100.

Waites G.T., James R.F.L. and Bell S.C. 1988b. "Human pregnancy associated α1-globulin," an insulin-like growth factor binding protein-immunohistological localization in the decidua and placenta during pregnancy employing monoclonal antibodies. J. Endocrinol. 120:351.

A ROLE FOR THE EXTRACELLULAR MATRIX IN AUTOCRINE AND PARACRINE REGULATION OF TISSUE-SPECIFIC FUNCTIONS

M. H. Barcellos-Hoff and
M. J. Bissell

Division of Cell and Molecular Biology
Lawrence Berkeley Laboratory
Berkeley, CA 94720

How can the knowledge gained in characterizing a functional epithelial cell model in culture shed light on the autocrine and paracrine mechanisms of gene regulation? Our purpose in presenting the following brief description of mammary epithelial cell culture is two-fold. First, to suggest that functional cells should be prerequisite for studying the effects and relationships between the cell and the multitude of growth factors and hormones that interact with tissues. Second, to emphasize the role of the extracellular matrix (ECM) in obtaining the "correct" functional phenotype for epithelial cells in culture. These aims are by no means all inclusive. There are many new and interesting reports of the effects of growth factors in mammary gland that we will not discuss in detail here, including exciting research that addresses how ECM and soluble factors interact *in vivo*. For example, the recent results indicating that transforming growth factor-β (TGF-β) acts as a negative regulator of ductal growth in virgin gland (Silbertstein and Daniel, 1987) may involve the ECM, since TGF-β appears to be intimately involved in synthesis and degradation of ECM molecules (Sporn *et al.*, 1987).

THE ROLE OF ECM IN MAMMARY EPITHELIAL DIFFERENTIATION

It has long been recognized that development and differentiation of organs *in vivo* depend on tissue interactions and changes in hormonal milieu that are accompanied by alterations in the ECM (Kratochwil, 1969; Sakakura *et al.*, 1976; Wessells, 1977; Topper and Freeman, 1980). In recent years it has become apparent that the extra-cellular environment is a crucial regulator of development and differentiation in cultured cells. As we gain knowledge of epithelial cell function and differentiation in culture, the model of dynamic reciprocity (Bissell *et al.*, 1982) gains further credence. In this model, it is postulated that the cell and ECM are inextricably bound together in a reciprocal relationship in which each influences the other via cytoskeleton connections to nuclear matrix and transmembrane bridges.

Based on current studies by this and other laboratories using a variety of tissues, we have formulated three principles that constitute a basis for studying the regulation of gene expression in epithelial tissues (Bissell and Hall, 1987): 1. The unit of function in multicellular organisms includes the cell and its extracellular matrix. This may also be extended to include cell-cell interactions, leading to the notion that in the last analysis the unit of function may be the entire organ itself. 2. The ECM-cell interaction is a continuum; that is, the ECM receives, imparts, and integrates structural and functional signals to the cell, which in turn modifies the ECM, which in turn modifies the cell, and so on. 3. That alterations in cell shape and structure are indications of the dynamics of this interaction, in as much as morphology reflects changes in cell polarity, cytoskeleton structure and altered organelle organization as the results of the cell's interaction with its environment. If cellular milieu is viewed as a fundamental determinant of differentiated gene expression, albeit via mechanisms that have only just begun to be elucidated, then tissue-specific gene regulation by autocrine and exocrine mechanisms may be modulated by the cell's interaction with its physical surroundings involving the cell membrane, cell contacts, and ECM (for an expaned review see Bissell and Hall, 1987).

It is now possible to culture functional epithelial cells. Thus, we have at hand the means to determine the relationships between various growth factors and hormones that influence an epithelial cell's specific pattern of gene expression and hormonal responsiveness. In this chapter we will briefly review culture strategies, developed by us and other researchers, for functional, differentiated mammary epitheial cells. Several aspects of this disscussion may be found in more detail in previous reviews, which are referenced as appropriate. In the context of this workshop on autocrine and exocrine mechanisms in reproductive endocrinology, it is our contention that relevant mechanisms of hormone action are most accurately elucidated in a "normal", functional and well-defined target cell population (Bissell, 1981).

Defining Differentiation In Culture

The success of obtaining functional differentiation in culture can be assessed by studying individual cell morphology to determine cytological differentiation, examining cytoarchitecture as an analog of morphogenesis, measuring response to mammotrophic hormones, and assaying milk protein, mRNA or DNA synthesis. Since at least two studies have demonstrated dissociation of cytological and functional differentiation in the mammary gland (Sakakura *et al.*, 1979; Vonderhaar and Smith, 1982), the most stringent protocols must assay each aspect of mammary epithelial differentiation before concluding that fully functional cells have been induced or maintained under culture conditions.

In these studies, it is essential to constantly refer back to the *in situ* tissue as a baseline for assessing culture conditions. Secretory epithelia are characterized both by their distinctive morphologies and by their specialized products. Pregnant and lactating mammary glands consist of epithelium organized into clusters of alveoli branching off of a ductal tree. During pregnancy, the epithelium lining the secretory alveoli of the gland differentiates into polarized columnar cells that contact a basement membrane at their basal surface and face the alveolar lumen at their apical surface. Cytodifferentiation is characterized by basally located nuclei, apically situated Golgi apparatus, secretory vesicles and fat droplets, and abundant apical microvilli. Shortly after partuition the lumen is sealed from the serosal space via tight junctions between epithelial cells. This facilitates transport of isotonic fluid and enhances vectorial compartmentalization of apically secreted milk proteins (Pitelka *et al.*, 1973; Neville, 1987). Polarized functional activities, such as the uptake of nutrients and of many receptor-mediated hormones at the basal epithelial surface and secretion of tissue-specific proteins at the apical surface, are an important characteristic of epithelia. Polarized secretion of milk proteins and compartmentalization are essential features of the functional lactating gland.

The relationship between cell replication and expression of tissue-specific function *in vivo* has been a matter of much controversy. DNA synthesis occurs *in vivo* primarily at the end buds during postnatal morphogenesis and pregnancy (Russo and Russo, 1980). The greatest growth potential resides in the putative stem cells, which have not been identified in culture, and myoepithelial cells (Dulbecco *et al.*, 1982). The physiological stimuli for growth is under the control of multiple hormones and growth factors that almost certainly involves epithelial-stromal interactions.

Mammary gland development is known to be dependent upon mesenchymal-epithelial interactions (Sakakura, 1987), and several investigators have confirmed this interaction with isolated cells. The mechanisms by which stromal cells influence, and are influenced by, epithelial cell differentiation have not been elucidated but probably involve, in part, elaboration of ECM by the stroma (Chiquet-Ehrismann *et al.*, 1986). Since the composition of the ECM of the mammary gland changes as a function of the stage of mammary gland growth and maturation (Silbertstein and Daniel, 1982; Warburton *et al.*, 1982), cellular proximity to specialized ECMs may be a primary factor during epithelial differentiation.

In particular, the thin, specialized basal lamina that separates epithelial cells from the stroma has been shown to change as a function of development and differentiation. The fully differentiated alveolar epithelium in lactating gland is in contact with a basement membrane composed of type IV collagen, laminin, fibronectin (Warburton *et al.*, 1982; Monaghan *et al.*, 1983), type V collagen, entactin (Warburton *et al.*, 1984), and sulfated glycosaminoglycans (Silbertstein and Daniel, 1982). Alterations in the basal lamina accompany physiological and pathological changes in many glandular tissues. During mammary gland involution some of the first changes occur in the basal lamina. It becomes thicker, is occasionally incomplete (Warburton *et al.*, 1982), and degradation of type IV collagen occurs (Martinez-Hernandez *et al.*, 1976). Mammary carcinoma is commonly associated with desmoplasia of the interstitial ECM and alterations in the basal lamina (Robbins and Cottran, 1979; Pitelka *et al.*, 1980; Ormerod and Rudland, 1985). In addition several studies have shown that normal gland development is disrupted when ECM deposition is inhibited (Wicha *et al.*, 1980; Silberstein and Daniel, 1982). From these data, and evidence that ECM profoundly influences milk protein synthesis and secretion in cultured cells (see below), we have concluded that basement membrane is a key regulator of mammary gland development and differentiation *in vivo* (Chen and Bissell, 1987).

CONSEQUENCES OF ECM-MAMMARY EPITHELIAL CELL INTERACTIONS

We have noted that a suitable substrata is often the pivotal element necessary for establishing culture conditions that elicit differentiated function in many cell types. In this section we will discuss studies that address how culture substrata influence murine mammary epithelial function. The details of our experimental protocol for isolating epithelial cells from mammary gland of midpregnant mice and culture

Table 1. **Mammary Epithelial Cell Isolation.**

Remove mammary glands, mince well.
Dissociate using collagenase digestion.
Use differential centrifugation to separate epithelial cells from fibroblasts.
Count and plate epithelial clusters and cells.
Culture with prolactin, hydrocortisone, and insulin plus serum for 2 days.
Remove serum after second day.

SUBSTRATA	PLASTIC	COLLAGEN GEL	EHS MATRIX
MORPHOLOGY	Flat	Cuboidal	Columnar
ARCHITECTURE	Monolayer	Monolayer, domes	Hollow spheres
FUNCTION	Low	High	Very high

conditions have been summarized recently (Bissell *et al.*, 1987; Emerman and Bissell, 1988) and are briefly summarized above.

Cellular Ultrastructure and Architecture

The relationship between substrata flexibility and cell shape changes has been studied in human, rabbit, rat and mouse mammary epithelial cells cultured on, or within, an attached or floating collagen gel (Emerman and Pitelka, 1977; Yang *et al.*, 1981; Richards *et al.*, 1983; Suard *et al.*, 1983). That cell-substratum interactions are significant regulators of tissue-specific behavior is apparent from the modulations in cell morphology on various substrata (Emerman and Pitelka, 1977; Li *et al.*, 1987; Bissell and Aggeler, 1987; Barcellos-Hoff *et al.*, 1989). The flat, unpolarized appearance of primary mammary epithelial cells cultured on conventional tissue culture plastic bears little resemblance to the columnar, polarized cells of the alveolar epithelium (Pitelka *et al.*, 1973; Emerman and Pitelka, 1977). Culturing cells on collagen gel produces a more differentiated appearance; the cells become polarized and increase the production of milk proteins over that obtained from plastic cultures. Floating the gel results in cell-induced gel contraction that leads to polarization of organelles, apical microvilli and formation of a basal lamina (Emerman and Pitelka, 1977; Burwen and Pitelka, 1980; Shannnon and Pitelka, 1981), and dramatic increases in milk protein synthesis and secretion (Lee *et al.*, 1984; Lee *et al.*, 1985; Blum *et al.*, 1987) and milk-specific mRNA levels (Lee *et al.*, 1985; Li *et al.*, 1987). Gluteraldhyde fixation of the gel blocks gel contraction upon floatation and suppresses these effects (Shannnon and Pitelka, 1981; Lee *et al.*, 1984). This requirement for substratum flexibility may be interpreted as allowing cells to assume a shape that is conducive to tissue-specific function, in this case, assumption of polarity and synthesis and secretion of milk proteins.

In contrast to cell culture configurations that result in an epithelial sheet, seeding small clusters and single cells onto a reconstituted basement membrane matrix (known as Matrigel or EHS for its derivation from the Engelbreth-Holm-Swarm murine sarcoma tumor) produces 3-dimensional "mammospheres" (Li *et al.*, 1987; Barcellos-Hoff *et al.*, 1987; Neville *et al.*, 1987; Barcellos-Hoff *et al.*, 1989). These differ from the domes found in monolayer cultures on plastic. The domes are the result of "blistering" due to fluid accumulation under the epithelial pavement (Pickett *et al.*, 1975), while multicellular structures organized

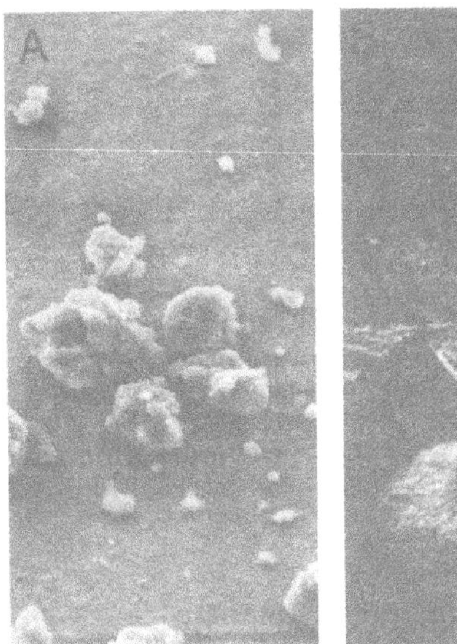

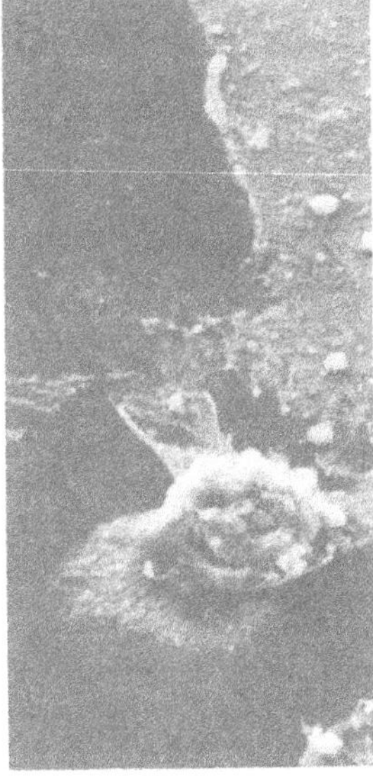

Figure 1. Scanning electron micrograph of mammary epithelial cells 3 hr (A) after plating on EHS matrix and 24 hr (B) later. Note that, in comparison to the solid matrix present in A, the cells have "pulled" on and reorganized the EHS matrix, as evidenced by the bare areas of plastic culture dish, during the first 24 hr in culture. x640 Original micrograph courtesy of Dr. J. Aggeler.

on EHS contain lumina completely enclosed by epithelial cells. The lumina appear forma via a process of cavitation within aggregates of cells (Barcellos-Hoff *et al.*, 1989). Scanning electron microscopy indicated that the cells reorganize the EHS matrix early in culture (Figure 1). Transmission microscopy of early cultures show solid clusters of cells with morphological evidence of necrotic cells within. As interior cells deteriorate, cells adjacent to the resulting cavity develop apical specializations. This results in the eventual formation of lumina faced with polarized epithelial cells that are joined with apical junctions and therefore appear capable of sequestering milk proteins (Figure 2). This sequestered compartment is analogous to the transjunctional permeability barrier in alveoli of lactating gland (Neville and Peaker, 1981). Thus, although the processes in culture may differ from those *in vivo*, the multicellular organization of mammary epithelial cells on EHS recapitulate the morphology of alveolar epithelia (Barcellos-Hoff *et al.*, 1989).

Cell Replication

It has been observed that different culture conditions elicit different levels of proliferation and functional differentiation; furthermore, one may be obtained without significant expression of the other (Haeuptle *et al.*, 1983; Suard *et al.*, 1983; Richards *et al.*, 1983; Lee *et al.*, 1984; Lee *et al.*, 1985; Wilde *et al.*, 1984). There is evidence that a round of DNA synthesis is prerequisite for differentiation in virgin mouse mammary gland explants (Smith and Vonderhaar, 1981). However, in other studies the role of replication is not so clear. Cells cultured on top of attached collagen gels proliferate but fail to differentiate morphologically (Suard *et al.*, 1983), while cells

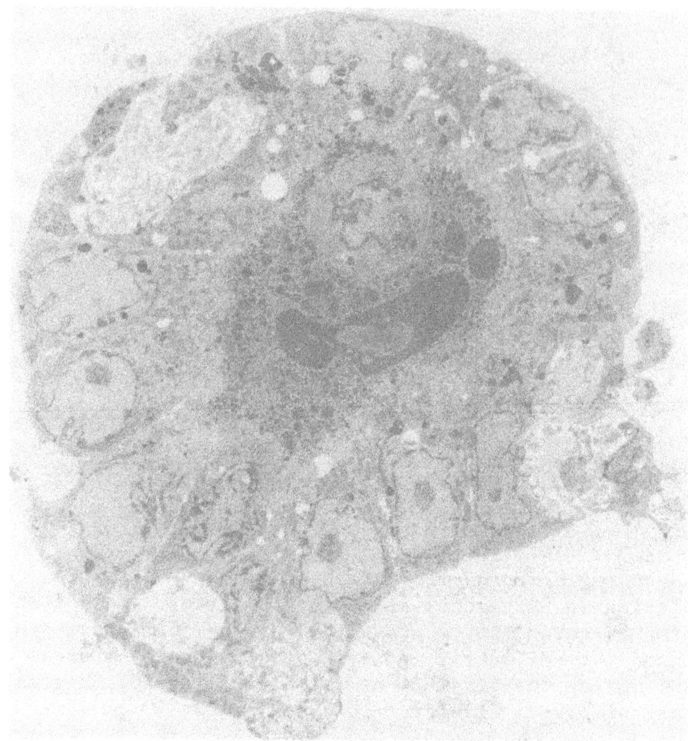

Figure 2. Ultrastructural architecture of mammary epithelial cells cultured for 4 days on EHS matrix. Transmission electron micrograph of a cross section through an aggregate of cells demonstrating lumina formation and the highly secretory cellular morphology. Note the abundant intraluminal secretions. x1700. Original micrograph courtesy of Dr. J. Aggeler.

on released gels do not grow but exhibit a greater degree of tissue-specific function and cytodifferentiation (Emerman *et al.*, 1977; Haeuptle *et al.*, 1983). In contrast, embedding cells within collagen gel results in simultaneous expression of both cell proliferation and milk protein synthesis (Haeuptle *et al.*, 1983; Suard *et al.*, 1983).

The relationship between replication and differentiation may be obscured by the pluripotent nature of the stem cell population of mammary gland (Dulbecco *et al.*, 1986), the heterogeneity of mammary gland epithelial isolations (Raber and D'Ambrosio, 1986) and/or different or conflicting nutritional requirements for each process (Flynn *et al.*, 1982; Imagawa *et al.*, 1985). The latter source of complexity is exemplified by the report of a two-step system in which primary mammary epithelial cells grow when embedded in an attached collagen gel and fed with serum- and EGF-containing medium and then differentiate when the gel is released and fed with serum-free medium (Flynn *et al.*, 1982). Whether growth and differentiation in mammary epithelial cells are sequential or concurrent in the same cell, or whether the differentiated cells in culture are due to selection of stem cell populations or subpopulation heterogeneity, remain to be determined but are essential problems in epithelial cell differentiation.

Tissue-Specific Function

The most characteristic feature of mammary gland is lactogenesis, in which the processes of growth, development and differentiation culminate in production of milk. Milk itself is a complex biological fluid consisting of lactose, lipids, vitamins, minerals, trace elements and proteins. In particular, skim milk proteins have been used to advantage as markers in studying the factors influencing functional differentiation under various culture conditions. Using the two-dimensional electrophoretic pattern of mouse skim milk proteins as reference, the secreted proteins of primary epithelial cells maintained on plastic, attached type I collagen gels or floating gels have been analyzed (Lee *et al.*, 1984). As expected, the pattern of secretion were found to be altered in cultured cells. Substratum- and hormone-induced changes in milk protein synthesis and secretion are complex and are not coordinated (Lee *et al.*, 1984; Lee *et al.*, 1985; Lee *et al.*, 1987). The non-coordinate expression of milk proteins in cultured cells underscores the experimental advantage of assaying function using the complex pattern of milk proteins. Since transferrin is expressed under all culture conditions, if we had chosen to study transferrin alone as a marker of function, we would have detected it under all culture conditions and perhaps have concluded that the culture conditions were equivalant. On the other hand, since α-lactalbumin is not abundant in culture, if it had been used as the marker of differentiated cells we may have concluded that the striking morphological changes indicative of secretory activity observed on collagen gels were not accompanied by tissue-specific function. It is also important to contrast not only the composition of the secretions obtained in culture but the relative proportion and total amount of each protein in relation to its pattern of expression *in vivo*, since there are physiologically relevant variations in ratios of milk proteins secreted *in vivo*. Although it is often encouraging to detect any tissue-specific markers in some culture systems, it is provocative to compare the levels in culture to those *in vivo*, particularly when the goal is to understand gene regulation. Finally it is informative to assess gene expression at multiple levels: mRNA transcription, processing and stability as well as protein synthesis, modification and secretion.

Using EHS as a substratum elicits increased expression of milk proteins and secretory activity compared to other substrata (Bissell *et al.*, 1985; Li *et al.*, 1987). Cells cultured on EHS secrete considerably more total protein than cells cultured on plastic and more secreted proteins are milk proteins (Figure 3). In addition, secretion is vectorial (preferentially into lumina) as shown by extracting sequestered proteins using a brief EGTA treatment to open cell junctions (Figure 3; Barcellos-Hoff *et al.*, 1989). This treatment produces no release of milk proteins from cells cultured on plastic. In comparison, EGTA extraction of cells cultured on EHS matrix produces a fraction enriched in milk proteins. Caseins are concentrated 2-8 fold in the EGTA extract over medium whereas transferrin and lactoferrin are present in comparable amounts in both compartments, suggesting that these proteins use different secretory pathways. Together with the alveolar morphology and columnar cytostructure, the polarized secretion of major milk proteins make this a suitable model for studying vectorial secretion and membrane domains.

Metabolite patterns of primary mammary epithelial cells are also a sensitive marker of the differentiated state (Bissell, 1981). The virgin gland does not synthesize glycogen, the late pregnant gland synthesizes appreciable levels, but at parturition glycogen breaks down and the ratio of lactose to glycogen increases dramatically (Emerman *et al.*, 1980). Cells from midpregnant mice maintained on plastic lose their distinctive

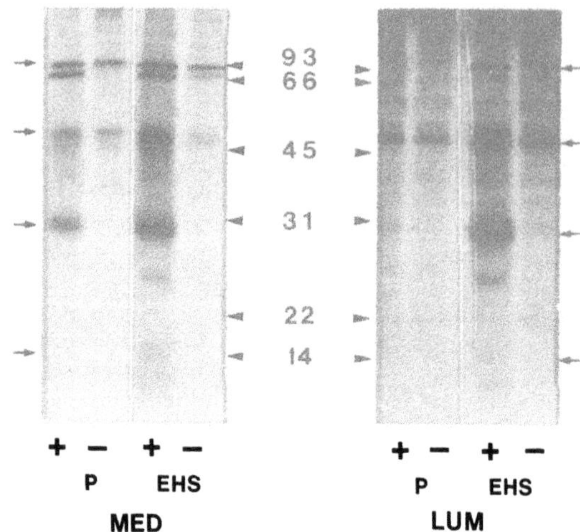

Figure 3. Milk protein secretion by primary mammary epithelial cells cutlured on plastic (P) or EHS matrix substrata in the presence (+) or absence (-) of lactogenic hormones. The proportional pattern of milk protein secretion into the medium (MED) or EGTA-sensitive (LUM) compartment extracted using 2.5 mM EGTA in Ca^{++}-free medium are shown. Samples were immunoprecipitated with a broad spectrum antibody to mouse skim milk and subjected to SDS-PAGE. Arrowheads indicate molecular weight markers (x 10^{-3}); small arrows indicate the location of abundant milk proteins.

metabolite patterns, while those on floating collagen gels do not (Emerman *et al.*, 1981). However, when cells from lactating mice are cultured on floating gels the metabolite pattern reverts to that characteristic of the pregnant state. Thus, the milieu of hormones, media and substrata used in this case may be more conducive to the expression of pregnant rather than the lactating phenotypes. It is also possible that interaction of epithelial and adipocyte components of the gland may be necessary to bring about the switch from glycogen to lactose synthesis after parturition.

Messenger RNA for milk proteins begin to accumulate during pregnancy and increase further during lactation (Hobbs *et al.*, 1982; Mercier and Gaye, 1983). β-casein mRNA expression is dependent upon the presence of prolactin, and mammary epithelial cells respond to the addition of prolactin by increasing the level of most milk protein messages (Hobbs *et al.*, 1982; Lee *et al.*, 1985; Chen and Bissell, 1987). The magnitude of the response in cultured cells is linked to the nature of the substrata; compared to cells on plastic, cells on floating gels are 5 times more responsive to prolactin (Lee *et al.*, 1985). In the presence of prolactin cells on floating collagen gels or EHS matrix produce more β-casein mRNA than the midpregnant gland from which they were derived (Figure 5A), and cells on EHS matrix can produce as much message for this protein as lactating gland (Li *et al.*, 1987). However individual components of basement membrane vary in their ability to maintain β-casein mRNA levels; type IV collagen and fibronectin are ineffective, while HSPG and laminin allow increased mRNA levels (Li *et al.*, 1987). Similar ECM effects have been observed for the COMMA-D cell strain (Bissell *et al.*, 1985; Medina *et al.*, 1987). Several investigators have reported data that indicates that substratum influences milk protein mRNA levels mainly at the post-transcriptional level (Rosen *et al.*, 1980; Blum *et al.*, 1987). The mechanism of this regulation is under investigation in our laboratory as well as others.

A further modification of culture environment is to provide a substrata of metabolically active, but reproductively dead cells that can alter the media by contributing their metabolic products, synthesize an ECM, and/or act directly by cell-cell interactions. Feeder layers of fibroblastic (Haslam, 1986), epithelial (Ehmann *et al.*, 1984) and mesenchymal (Levine and Stockdale, 1984; Levine and Stockdale, 1985; Wiens *et al.*, 1987) origin have been used as substrata for mammary epithelial growth and differentiation, many of which promote the deposition of a basal lamina. Growth and formation of a ductal-like morphology occurs when mammary epithelial cells are cultured with proadipocytes, independently of lactogenic hormones, while cytodifferentiation and milk protein production are hormone-dependent (Wiens *et al.*, 1987). Stromal-epithelial interactions have particular importance in development and recent studies have shown that some mechanisms of tissue-regulation may be linked in these two cell types (see Cunha, Chapter). Haslam has used homologous mammary fibroblasts from mammary epithelial cell isolations to determine the role of stroma in mammary gland growth and differentiation. These studies have shown that fibroblasts influence the response of isolated mammary epithelial cells to estrogen and their regulation of progesterone receptors (Haslam, 1986; Haslam and Levely, 1985).

INTERACTIONS RELATING TISSUE-SPECIFIC FUNCTION AND ECM

Studies of the interaction of mammary epithelial cells with ECM and/or its components in culture underscore a physiological role for ECM *in vivo*. Acellular biomatrices extracted from mammary gland, but not those from liver (Wicha *et al.*, 1982), have been shown to support long term growth and differentiation (Wilde *et al.*, 1984). Primary mammary epithelial cells preferentially attach to, and proliferate on, collagen type IV compared to type I, with an enhancement in growth if laminin is added (Wicha, 1984). EHS-reconstituted basement membrane supports cytological differentiation and high levels of milk protein synthesis and secretion, but individual basement membrane components by and large do not (Bissell *et al.*, 1985; Li *et al.*, 1987; Medina *et al.*, 1987; Blum *et al.*, 1987).

Mammary epithelial cells on floating collagen gels and EHS, but not on plastic or attached gels, have been shown to produce a continuous basement membrane (Emerman and Pitelka, 1977; Richards *et al.*, 1982; Bissell and Aggeler, 1987; Barcellos-Hoff *et al.*, 1988). The demonstration of an organized basal lamina is additional evidence of functional polarity in these cells and is another consequence of the cell-substratum interaction. The synthesis of ECM components by cultured cells also changes in response to substratum (David and Bernfield, 1979; Parry *et al.*, 1982; Parry *et al.*, 1985; David *et al.*, 1987). One mechanism by which collagen substrata appear to promote basal lamina formation is by stabilizing incorporation of secreted ECM molecules into a basal lamina. Parry and collegues found that most of the glycosaminoglycans synthesized by cells on plastic are secreted into the medium, while those secreted by cells on floating collagen gels are incorporated into the ECM (Parry *et al.*, 1985).

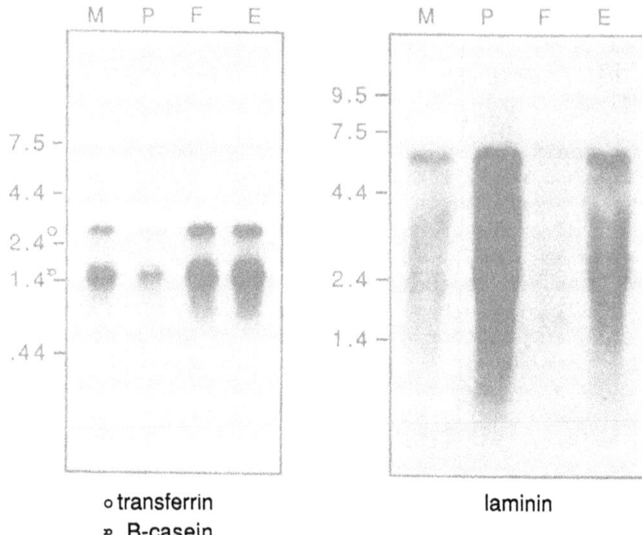

Figure 4. Expression of tissue-specific mRNAs of mammary epithelial cells *in vivo* and cultured on various substrata. A. Double northern for β-casein and transferrin of 2 μg total mRNA from mammary gland (M) obtained at midpregnancy, or from cells cultured for 6 days on plastic (P), floating collagen gel (F) or EHS matrix (E) substrata. B. Northern for laminin of 10 μg total mRNA obtained from cells as in A. Quantitation by scanning densitometry of slot blots hybridized in parallel indicates that ratio of the amount of laminin mRNA obtained from midpregnant gland to that of the cells on plastic was 3, from cells on floating collagen gel was 0.39 and from cells cultured on EHS was 1.4. The positions of RNA size markers (kB) are shown in the margin.

Since culture configurations that produce differentiated function also elicit a distinct basal lamina, functional changes in these cultures may be effected through the influence of newly synthesized basement membrane (Bissell and Hall 1987). The levels of mRNA for ECM proteins are regulated in the mammary gland by hormonal status and in cultured epithelial cells by substratum, which may be related to the changes in basement membrane protein deposition seen morphologically (Figure 4B; Streuli *et al.*, 1988). A role for ECM occurs in normal gland is indicated by the disruption of normal gland development when ECM deposition is inhibited (Wicha et al., 1980; Silberstein and Daniel 1982a) as well as by the observation that an early event in gland involution is disassembly of the basement membrane (Martinez-Hernandez *et al.*, 1976). Thus, modulation of function by substrata may in part be due to its effect on the secretion and organization of a tissue-specific basement membrane, which in turn could affect mammary cell differentiation. Whether endogenously produced ECM is a primary event in the functional differentiation of mammary cells is currently under investigation in our laboratory (Streuli *et al.*, submitted).

In general, a change in cytostructure and polarity appears to accompany ECM-induced functional changes. This influence on cell shape may be the result of a direct interaction between ECM proteins and cell surface proteins that then serve to anchor structural elements so that the cell can generate its own tensile forces (Ingber *et al.*, 1986). Whether such morphological changes are a prerequisite for tissue-specific function remains to be determined. However, it is well documented that cell spreading is associated with proliferation (Ormerod and Rudland, 1985) whereas a polarized cytoplasm and a more columnar shape is related to epithelial differentiation (Watt, 1986). Important experiments by Penman and colleagues have shown that mRNA metabolism is drastically altered by the shape changes accompanying cell suspension (Penman *et al.*, 1981). The polarized epithelial cell shape, as opposed to the flattened cell morphology (Emerman and Pitelka, 1977), does appear to be necessary for maintaining significant levels of milk proteins mRNA, synthesis and secretion (Bissell *et al.*, 1982; Haeuptle *et al.*, 1983; Flynn *et al.*, 1982; Lee *et al.*, 1984; Lee *et al.*, 1985; Barcellos-Hoff *et al.*, 1989).

Multicellular interactions are integral to the form and pattern of tissues and those in mammary gland that result in lobuloalveolar morphogenesis are a prominant consequence of differentiation (Daniel and Silberstein, 1987). In mammary culture, EHS matrix has a unique capacity to promote histiotypic morphology in mammary epithelial cells (Bissell *et al.*, 1985; Li *et al.*, 1987; Medina *et al.*, 1987), although lumen formation is also seen when cells are embedded (Haeuptle *et al.*, 1983; Flynn *et al.*, 1982; Ormerod and Rudland, 1988) or overlayed with collagen (Hall *et al.*, 1982). A remarkable aspect of the use of EHS matrix as a culture substrata is that it produces enhanced differentiated function of Sertoli cells (Hadley *et al.*, 1985), hepatocytes (Bissell *et al.*, 1987), human endometrial secretory gland (Rinehart *et al.*, 1987), lung alveolar epithelial cells (Shannon *et al.*, 1987; Blau *et al.*, 1988) and keratinocytes (Baskin *et al.*, 1987) and undoubtedly other tissues. EHS also leads to reorganization--or 'morphogenesis'-- into a pattern specific to the tissue of origin: hollow, alveoli-like spheres for lung and mammary epithelial cells, trabecular cords for hepatocytes, and polarized tubes for Sertoli cells. The mechanism by which EHS exerts these influences on epithelial cell cytostructure, morphology and gene expression has yet to be elucidated. We have proposed a model, shown in Figure 5, that postulates that cell contact with ECM, possibly via receptors such as the integrins, imparts directionality to a cell (Barcellos-Hoff *et al.*, 1989). This leads to the establishment of polarity that promotes the expression of tissue-specific polarized secretion.

The ability of ECM to bind growth factors and serve as a "sink" or reservoir is also gaining prominence as a component of the role of ECM in differentiation and development. For instance, endothelial cells incorporate highly stable b fibroblast growth factor into their ECM (Vlodavsky *et al.*, 1987), which suggests that the influence of sequestered growth factors may be difficult to distinguish from the influences of the ECM molecules themselves. These data are consistant with a physiological role for ECM *in vivo* that may involve concentration and "presentation" of growth factors.

The role of growth factors in inducing changes in ECM protein production and deposition have just begun to be investigated. As in other cellular responses to growth factors, it is a complex effect involving mechanisms at multiple levels. This complexity is illustrated in a recent study using primary mammary epithelial cultures, indicating that growth factors that act via the EGF receptor (TGFα and EGF) stimulate these cells to make collagen type IV when the cells are plated onto a collagen gel but not when they are on collagen type IV. This appears to effected through a substratum modulated change in the number of surface receptors for these factors (Mohanam *et al.*, 1988). Mammary epithelial cell also synthesize transforming growth factor α, and its synthesis is substratum and hormone sensitive (Zwiebel *et al.*, 1982; Liu *et al.*, 1987), suggesting that TGFα may have a role in the deposition of basement membrane *in vivo*.

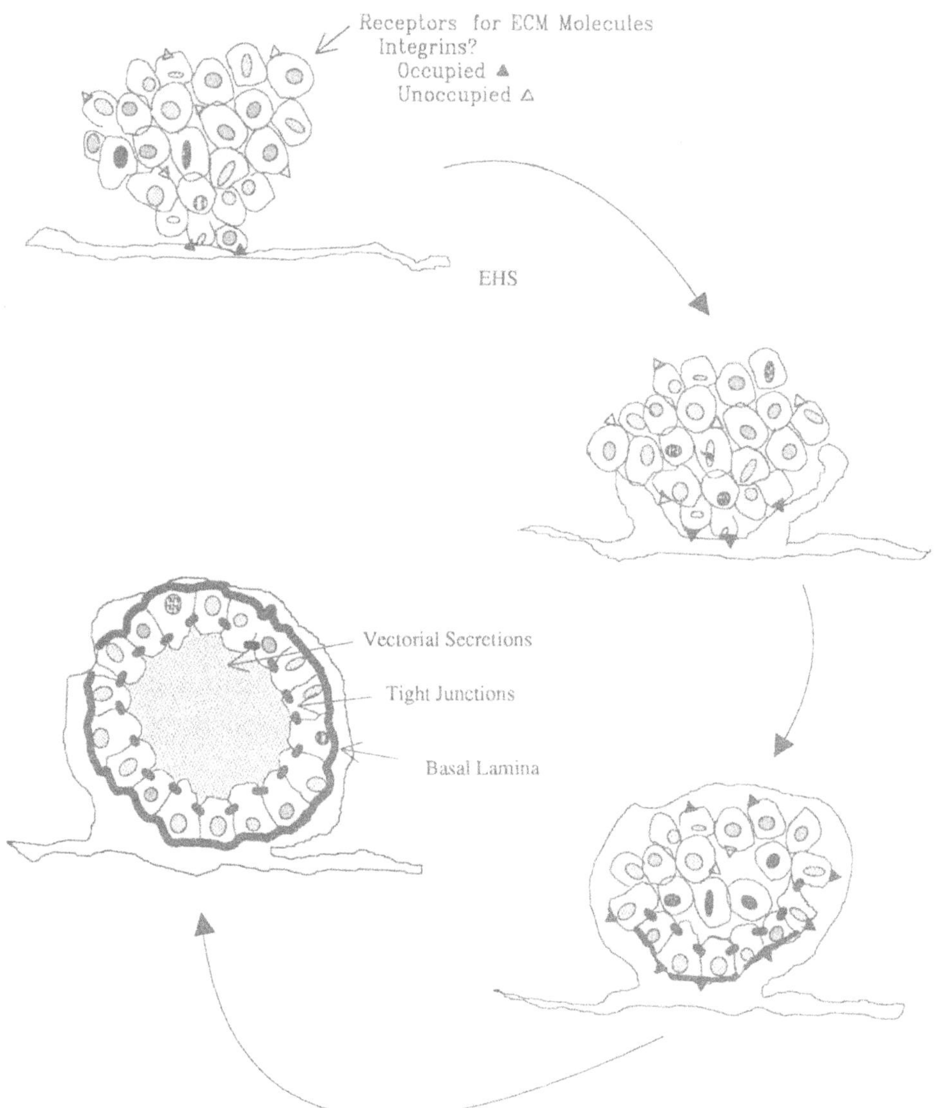

Figure 5. Model of alveolar-like morphogenesis by mammary epithelial cells cultured on EHS matrix.

In this chapter, we have summarized a body of data indicating that physiologically relevant criteria for differentiation in cultured epithelial cells provide the most rigourous baseline for studying the regulation of gene expression by hormones. The role of autocrine and paracrine factors in this system, as well as *in vivo*, have just begun to be elucidated but are beyond the scope of this paper. The perspective that the cell is a continuum with its environment, that each influences the other via dynamic interactions, will hopefully provide new insight for researchers exploring the regulatory functions of soluable factors.

REFERENCES

Barcellos-Hoff M.H., Neville P.N., Aggeler J. and Bissell M.J. 1987. Polarized secretion by mammary epithelial cell cultures on EHS-matrix. J. Cell Biol. 105:220a.

Barcellos-Hoff M.H., Aggeler J., Ram T.G. and Bissell M.J. 1989. Functional differentiation and alveolar morphogenesis of primary mammary epithelial cells cultures on reconstituted basement membrane. Development 105:225.

Baskin J.B., Shannon J.M. and Clark R.A.F. 1987. Early events in the multicellular organization of human epidermal cells on basement membrane matrix. J. Cell Biol. 105:223a.

Bissell D.M., Arenson D.M., Maher J.J. and Roll F.J. 1987. Support of cultured hepatocytes by a laminin-rich gel. J. Clin. Invest. 79:801.

Bissell M.J. 1981. The differentiated state of normal and malignant cells or how to define a "normal" cell in culture. Int. Rev. Cytol. 70:27.

Bissell M.J., Hall H.G. and Parry G. 1982. How does the extracellular matrix direct gene expression?. J. Theor. Biol. 99:31.

Bissell M.J. and Hall H.G. 1987. Form and function in the mammary gland: The role of extracellular matrix. In: "The mammary gland: development, regulation and function" (M. Neville and C. Daniel, eds.), Plenum Press, New York. p. 97.

Bissell M.J., Lee EY-H., Li M-L., Chen L-H. and Hall H.G. 1985. Role of extracellular matrix and hormones in modulation of tissue-specific functions in culture: mammary gland as a model for endocrine sensitive tissues. In: "Benign prostatic hyperplasia, Vol. II" (C.H. Rogers, D.C. Coffey, G.R. Cunha, eds.), NIH Publication No. 87-2881, Washington, D.C. p. 39.

Bissell M.J. and Aggeler J. 1987. Dynamic reciprocity: How do extracellular matrix and hormones direct gene expression? *In*: "Mechanisms of signal transduction by hormones and growth factors" (M. Cabot, ed.), Alan R. Liss, New York. p. 251.

Bissell M.J., Li M.L., Chen L-H. and Lee EY-H. 1987. Regulation of milk proteins in the mouse mammary epithelial cells by extracellular matrix and hormones. In: "Growth and differentiation of mammary epithelial cells in culture" (J. Enami and R. Ham, eds.), Scientific Society Press, Tokyo, Japan. p. 155.

Blau H., Guzowski D.E., Siddiqi Z.A., Scarpelli E.M. and Bienkowski R.S. 1988. Fetal type 2 pneumocytes form alveolar-like structures and maintain long-term differentiation on extracellular matrix. J. Cell. Physiol. 136:203.

Blum J.L., Zeigler M.E. and Wicha M.S. 1987. Regulation of rat mammary gene expression by extracellular components. Expt. Cell Res. 173:322.

Burwen S.J. and Pitelka D.R. 1980. Secretory function of lactating mouse mammary epithelial cells cultured on collagen gels. Expt. Cell Res. 126:249.

Chen L-H. and Bissell M.J. 1987. Transferrin mRNA level in the mouse mammary gland is regulated by pregnancy and extracellular matrix. J. Biol. Chem. 262:17247.

Chiquet-Ehrismann R., Mackie E.J., Pearson C.A. and Sakakura T. 1986. Tenascin: An extracellualr matrix protein involved in tissue interactions during fetal development and oncogenesis. Cell 47:131.

Daniel C.W. and Silberstein G.B. 1987. Postnatal development of the rodent mammary gland. In: "The mammary gland: development, regulation and function," (M. Neville and C. Daniel, eds.), Plenum Press Publishing Corp., New York. p. 3.

David G. and Bernfield M. 1979. Collagen reduces glycosaminoglycan degradation by cultured mammary epithelial cells: Possible mechanism for basal lamina formation. Proc. Natl. Acad. Sci. (USA) 76:786.

David G., Nusgens B., van der Schueren B., Cauwenberge D.V., van der Berghe B. and Lapiere C. 1987. Collagen metabolism and basement membrane formation in cultures of mouse mammary epithelial cells. Expt. Cell Res. 170:402.

Dulbecco R., Henahan M. and Armstrong B. 1982. Cell types and morphogenesis in the mammary gland. Proc. Natl. Acad. Sci. (USA) 79:7346.

Dulbecco R., Allen W.R., Bologna M. and Bowman M. 1986. Marker evolution during the development of the rat mammary gland: stem cells identified by markers and the role of myoepithelial cells. Cancer Research 46:2449.

Ehmann U.K., Peterson W.D.Jr. and Misfeldt D.S. 1984. To grow mouse mammary epithelial cells in culture. J. Cell Biol. 98:1026.

Emerman J.T. and Bissell M.J. 1988. Cultures of mammary epithelial cells: Extracellular matrix and functional differentiation. In: "Advances in cell culture," (K. Maramorosch and G.H. Sato, eds.), Academic Press, San Diego, CA. p. 137.

Emerman J.T., Bartley J.C. and Bissell M.J. 1980. Interrelationship of glycogen metabolism and lactose synthesis in mammary epithelial cells of mice. Biochem J. 192:695.

Emerman J.T. and Pitelka D.R. 1977. Maintenance and induction of morphological differentiation in dissociated mammary epithelium on floating collagen membranes. *In Vitro* 13:316.

Emerman J.T., Enami J., Pitelka D.R. and Nandi S. 1977. Hormonal effects on intracellular and secreted casein in cultures of mouse mammary epithelial cells on floating collagen membranes. Proc. Natl. Acad. Sci. (USA) 74:4466.

Emerman J.T., Bartley J.C. and Bissell M.J. 1981. Glucose metabolite patterns as markers of functional differentiation in freshly isolated and cultured mouse mammary epithelial cells. Expt. Cell Res. 134:241.

Flynn S., Yang J. and Nandi S. 1982. Growth and differentiation of primary cultures of mouse mammary epithelium embedded in collagen gel. Differentiation 22:191.

Hadley M.A., Byers S.W., Suarez-Quian C.A., Kleinman H. and Dym M. 1985. Extracellular matrix regulates Sertoli cell differentiation, testicular cord formation, and germ cell development *in vitro*. J. Cell Biol. 101:1511.

Haeuptle M-T., Suard Y.L.M., Bogenmann E., Reggio H., Racine L. and Kraehenbuhl J-P. 1983. Effect of cell shape change on the function and differentiation of rabbit mammary cells in culture. J. Cell Biol. 96:1425.

Hall H.G., Farson D.A. and Bissell M.J. 1982. Lumen formation by epithelial cell lines in response to collagen overlay: A morphogenetic model in culture. Proc. Natl. Acad. Sci. (USA) 79:4672.

Haslam S.Z. and Levely M.L. 1985. Estrogen responsiveness of normal mouse mammary cells in primary cell culture: Association of mammary fibroblasts with estrogenic regulation of progesterone receptors. Endocrinology 116:1835.

Haslam S.Z. 1986. Mammary fibroblast influence on normal mouse mammary epithelial cell responses to estrogen *in vitro*. Cancer Research 46:310.

Hobbs A.A., Richards D.A., Kessler D.J. and Rosen J.M. 1982. Complex hormonal regulation of rat casein gene expression. J. Biol. Chem. 257:3598.

Imagawa W., Tomocka Y., Hamamoto S. and Nandi S. 1985. Stimulation of mammary epithelial cell growth *in vitro*: Interaction of epidermal growth factor and mammogenic hormones. Endocrinology 116:1514.

Ingber D.E., Madri J.A. and Jamieson D.J. 1986. Basement membrane as a spatial organizer of polarized epithelia. Amer. J. Pathol. 122:129.

Kratochwil K. 1969. Organ specificity in mesenchymal induction demonstrated in the embryonic development of the mammary gland of the mouse. Dev. Biol. 20:46.

Lee EY-H., Parry G. and Bissell M.J. 1984. Modulation of secreted proteins of mouse mammary epithelial cells by the extracellular matrix. J. Cell Biol. 98:146.

Lee EY-H., Lee W-H., Kaetzel C.S., Parry G. and Bissell M.J. 1985. Interaction of mouse mammary epithelial cells with collagenous substrata: Regulation of casein gene expression and secretion. Proc. Natl. Acad. Sci. (USA) 82:1419.

Lee EY-H., Barcellos-Hoff M.H., Chen L-H., Parry G. and Bissell M.J. 1987. Transferrin is a major mouse milk protein and is synthesized by mammary epithelial cells. *In Vitro* Cell Dev. Biol. 23:221.

Levine J.F. and Stockdale F.E. 1984. 3T3-L1 adipocytes promote the growth of mammary epithelial. Expt. Cell Res. 151:112.

Levine J.F. and Stockdale F.E. 1985. Cell-cell interations promote mammary epithelial cell differentiation. J. Cell Biol. 100:1415.

Li M.L., Aggeler J., Farson D.A., Hatier C., Hassell J. and Bissell M.J. 1987. Influence of a reconstituted basement membrane and its components on casein gene expession and secretion in mouse mammary epithelial cells. Proc. Natl. Acad. Sci. (USA) 84:136.

Liu S., Sanfilippo B., Perroteau I., Derynck R., Salomon S.S. and Kidwell W.R. 1987. Expression of transforming growth factor alpha in differentiated rat mammary tumors: estrogen induction of transforming growth factor alpha production. Mol. Endocrinol. 1:683.

Martinez-Hernandez A., Fink L.M. and Pierce G.B. 1976. Removal of basement membrane in the involuting breast. Lab. Invest. 34:455.

Medina D., Li M.L., Oborn C.J. and Bissell M.J. 1987. Casein gene expression in mouse mammary epithelial cell lines: dependence upon extracellular matrix and cell type. Expt. Cell Res. 172:192.

Mercier J-C. and Gaye P. 1983. Milk protein synthesis. In: "Biochemistry of lactation," (T.B. Mepham, ed.), Elsevier Science Publishers, Amsterdam. p. 177.

Mohanam S., Salomon D.S. and Kidwell W.R. 1988. Substratum modulation of epidermal growth factor receptor expression by normal mouse mammary cells. J. Dairy Sci. 71:1507.

Monaghan P., Warburton M.J., Perusinghe N. and Rudland P.S. 1983. Topographical arrangement of basement membrane proteins in lactating rat mammary gland: Comparison of the distribution of type IV collagen, laminin, fibronectin, and Thy-1 at the ultrasturctual level. Proc. Natl. Acad. Sci. (USA) 80:3344.

Neville M.C. and Peaker M. 1981. Ionized calcium in milk and the integrity of the mammary epithelium in the goat. J. Physiol. 313:561.

Neville M.C., Stahl L., Lutes C., Bissell M.J. and Barcellos-Hoff M.H. 1987. A scanning electron microscope study of the morphogenesis of mouse mammary cultures on collagen gels and EHS matrix. J. Cell Biol. 105:139a.

Ormerod E.J. and Rudland P.S. 1985. Isolation and differentiation of cloned epithelial cell lines form normal rat mammary glands. *In Vitro* Cell Dev. Biol. 21:143.

Ormerod E.J. and Rudland P.S. 1988. Mammary gland morphogenesis *in vitro*: Extracellular requirements for the formation of tubules in collagen gels by a cloned rat mammary epithelial cell line. *In Vitro* Cell Dev. Biol. 24:17.

Parry G., Lee EY-H. and Bissell M.J. 1982. Modulation of the differentiated phenotype of cultured mouse mammary epithelial cells by collagen substrata. In: "The extracellular matrix," (S.P. Hawkes and G. Wang, eds.), Academic Press, New York. p. 303.

Parry G., Lee EY-H., Farson D.A., Koval N. and Bissell M.J. 1985. Collagenous substrata regulate the nature and distribution of glycosaminoglycans produced by differentiated cultures of mouse mammary epithelial cells. Expt. Cell Res. 156:487.

Penman S., Fulton A., Capco D., Ben-Ze'ev A., Willtelsberger S. and Tse C.F. 1981. Cytoplasmic and nuclear architechture in cells and tissue: Form, function and mode of assembly. Cold Spring Harbor Symp. Quant. Biol. 46:1013.

Pickett P.B., Pitelka D.R., Hamamoto S.T. and Misfeldt D.S. 1975. Occluding junctions and cell behavior in primary cultures of normal and neoplastic mammary gland cells. J. Cell Biol. 66:316.

Pitelka D.R., Hamamoto S.T., Duafala J.G. and Nemanic M.K. 1973. Cell contacts in the mouse mammary gland. J. Cell Biol. 56:797.

Pitelka D.R., Hamamoto S.T. and Taggart B.N. 1980. Basal lamina and tissue recognition in malignant mammary tumors. Cancer Research 40:1600.

Raber J.M. and D'Ambrosio S.M. 1986. Isolation of single cell suspensions from the rat mammary gland: Separation, characterization, and primary culture of various cell populations. *In Vitro* Cell Dev. Biol. 22:429.

Richards J., Guzman R., Konrad M., Yang J. and Nandi S. 1982. Growth of mouse mammary gland end buds cultured in a collagen gel matrix. Expt. Cell Res. 141:433.

Richards J., Pasco D., Yang J., Guzman R. and Nandi S. 1983. Comparison of the growth of normal and neoplastic mouse mammary cells on plastic, on collagen gels and in collagen gels. Expt. Cell Res. 146:1.

Rinehart C.A., Irigaray M.F., Lyn-Cook B.D. and Kaufman D.G. 1987. An *in vitro* model system for the organogenesis of human edometrial secretory glands. J. Cell Biol. 105:42a.

Robbins S.L. and Cottran R.S. 1979. Pathologic basis of disease. W.B. Sanders Co., Philadelphia

Rosen J.M., Matusik R.J., Richards D.A., Gupt P. and Rodgers J.R. 1980. Multihormonal regulation of casein gene expression at the transcriptional and post-transcriptional levels in the mammary gland. Recent Prog. Horm. Res. 36:157.

Russo J. and Russo I.H. 1980. Influence of differentiation and cell kinetics on the susceptibility of the rat mammary gland to carcinogenesis. Cancer Research 40:2677.

Sakakura T., Nishizuka Y. and Dawe C.J. 1976. Mesenchyme-dependent morphogenesis and epithelium-specific cytodifferentiation in mouse mammary gland. Science 194:1439.

Sakakura T., Sakagami Y. and Nishizuka Y. 1979. Persistence of responsiveness of adult mouse mammary gland to induction by embryonic mesenchyme. Dev. Biol. 72:201.

Sakakura T. 1987. Mammary embryogenesis. In: "The mammary gland: development, regulation and function," (M. Neville and C.W. Daniel, eds.), Plenum Press, New York. p. 37.

Shannnon J.M. and Pitelka D.R. 1981. The influence of cell shape on the induction of functional differentiation in mouse mammary cells *in vitro*. *In Vitro* 17:1016.

Shannon J.M., Mason R.J. and Jennings S.D. 1987. Functional differentiation of alveolar type II epithelial cells *in vitro*: effects of cell shape, cell-matirx interactions and cell-cell interactions. Biochem. Biophys. Acta 931:143.

Silberstein G.B. and Daniel C.W. 1982. Elvax 40P implants, sustained local release of bioactive molecules influencing the mammary ductal development. Dev. Biol. 93:272.

Silbertstein G.B. and Daniel C.W. 1982. Glycosaminoglycans in the basal lamina and extracellular matrix of the developing mouse mammary duct. Dev. Biol. 90:215.

Silbertstein G.B. and Daniel C.W. 1987. Reversible inhibition of mammary gland growth by transforming growth factor-β. Science 237:291.

Smith G.H. and Vonderhaar B.K. 1981. Functional differentiation in mouse mammary gland epithelium is attained through DNA synthesis, inconsequent of mitosis. Dev. Biol. 88:167.

Sporn M.B., Roberts A.B, Wakefield L.M. and de Crombrugghe B. 1987. Some recent advances in the chemistry and biology of transforming growth factor-β. J. Cell Biol. 105:1039.

Streuli C.H., Ram T.G. and Bissell, M.J. 1988. Basement membrane involvement in mammary gland differentiation. J. Cell Biol. 107:663a.

Streuli C.H., Ram T.G. and Bissell M.J. 1989. Expression of extracellular matrix components is regulated by substrata. Submitted.

Suard Y.M.L., Haeuptle M-T., Fairnon E. and Kraehenbuhl J-P. 1983. Cell proliferation and milk protein gene expression in rabbit mammary cell cultures. J. Cell Biol. 96:1435.

Topper Y.J. and Freeman C.S. 1980. Multiple hormone interactions in the developmental biology of the mammary gland. Physiol. Rev. 60:1049.

Vlodavsky I., Folkman J., Sullivan R., Fridman R., Ishai-Michaile R., Sasse J. and Klagsbrun M. 1987. Endothelial cell-derived basic fibroblast growth factor: Synthesis and deposition into subendothelial extracellular matrix. Proc. Natl. Acad. Sci. (USA) 84:2292.

Vonderhaar B.K. and Smith G.H. 1982. Dissociation of cytological and functional differential in virgin mouse mammary gland during inhibition of DNA synthesis. J. Cell Biol. 53:97.

Warburton M.J., Mitchell D., Ormerod E.J. and Rudland P. 1982. Distribution of myoepithelial cells and basement membrane proteins in the resting pregnant, lactating and involuting rat mammary gland. J. Histochem. Cytochem. 30:667.

Warburton M.J., Monoghan P., Ferns S.A., Rudland P.S., Perusinghe N. and Chung A.E. 1984. Distribution of entactin in the basement membrane of the rat mammary gland. Expt. Cell Res. 152:240.

Watt F.M. 1986. The extracellular matrix and cell shape. TIBS 11:482.

Wessells N.K. 1977. "Tissue interactions and development." Benjamin/Cummings, California

Wicha M.S., Liotta L.A., Vonderhaar B.K. and Kidwell W.R. 1980. Effects of inhibition of basement membrane collagen deposition on rat mammary gland development. Dev. Biol. 80:253.

Wicha M.S., Lowrie G., Kohn E., Bagavandoss P. and Mahn T. 1982. Extracellular matrix promotes mammary epithelial growth and differentation *in vitro*. Proc. Natl. Acad. Sci. (USA) 79:3213.

Wicha M.S. 1984. Interaction of rat mammary epithelium with extracellular matrix components. In: "New approaches to the study of benign prostatic hyperplasia," (F.A. Kimball, A.E. Buhl and D.B. Canter, eds.), Alan R. Liss, New York. p. 129.

Wiens D., Park C.S. and Stockdale F.E. 1987. Milk protein expression and ductal morphogenesis in the mammary gland *in vitro*: Hormone-dependent and -independent phases of adipocyte-mammary epithelial cell interaction. Dev. Biol. 120:245.

Wilde C.J., Hasan H.R. and Mayer R.J. 1984. Comparison of collagen gels and mammary extracellular matrix as substrata for study of terminal differentiation in rabbit mammary epithelial cells. Expt. Cell Res. 151:519.

Yang J., Elias J.J., Petrakis N.L., Wellings S.R. and Nandi S. 1981. Effects of hormones and growth factors on human mammary epithelial cells in collagen gel culture. Cancer Research 41:1021.

Zwiebel J.A., Davis M., Kohn E., Salomon D.S. and Kidwell W.R. 1982. Anchorage-independent growth-conferring factor production by rat mammary tumor cells. Cancer Research 42:5117.

A NEW CONCEPT OF BREAST CANCER GROWTH REGULATION AND ITS POTENTIAL CLINICAL APPLICATIONS

C. Kent Osborne

Department of Medicine/Oncology
University of Texas Health Science Center
San Antonio, TX 78284-7884

Major changes in our conceptual thinking about the regulation of growth of human breast cancer have occurred in the past decade. In the early 1970s our understanding of the mechanisms regulating breast cancer cell proliferation was simplistic. The pituitary hormone prolactin was thought to be important due to its critical involvement in normal breast differentiation and function and because it has a pivotal role in rodent breast cancer (Welsch and Nagasawa, 1977). However, the importance of this hormone in human breast cancer remains to be defined. The female sex steroid hormone estrogen, on the other hand, was known to be an important growth factor since the observation that about one-third of patients with breast cancer would have temporary tumor regression following treatments designed either to lower the serum estrogen level or to block its effects. Later estrogens and antiestrogens also were reported to have direct effects on proliferation of cultured human breast cancer cells (Lippman *et al.*, 1977). At this time cumulative data suggested that the mechanisms of estrogen action on cells involved the binding of estrogen to the estrogen receptor protein followed by the tight coupling of the receptor-hormone complex to DNA resulting in alterations of specific gene transcription and protein synthesis. Exactly how estrogens increased cell proliferation was not known. Nevertheless, knowledge of these simple biochemical pathways had important clinical implications that eventually led to the use of estrogen receptor assays for predicting response to endocrine therapy and for predicting the clinical course of patients with early stage disease.

This simplistic view of breast cancer growth regulation contrasts sharply with a model derived from more recent laboratory studies (Figure 1). This model maintains the previous estrogen response pathway mediated through cellular estrogen receptors. However, in addition it includes the possibility that, *in vivo*, estrogen could indirectly influence tumor growth through the secretion of growth factors (estromedins) produced at distant sites (Ikeda *et al.*, 1983). Evidence is also mounting that, in addition to steroid hormones, a host of polypeptide hormones and growth factors may play a role in the regulation of breast cancer growth.

Twelve years ago we observed that certain cultured human breast cancer cell lines were responsive to physiologic concentrations of insulin and that breast cancer cells contain specific, high-affinity

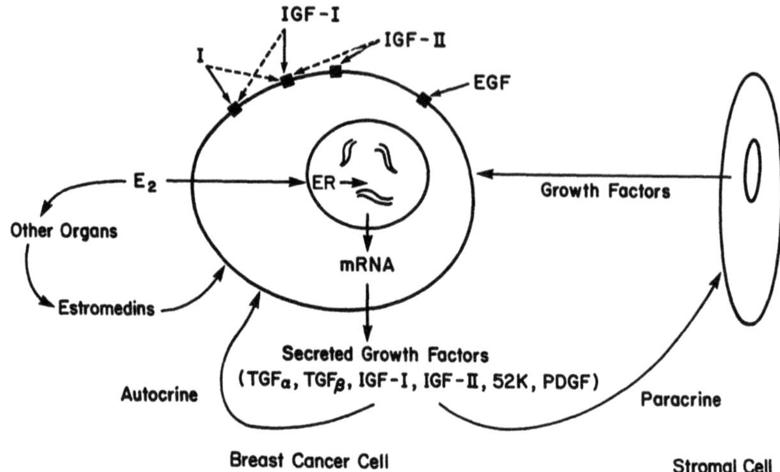

Figure 1. A model of breast cancer growth regulation illustrating potential autocrine and paracrine interactions between estrogen and growth factor.

membrane insulin receptors (Osborne *et al.*, 1976; Osborne *et al.*, 1978). Later, receptors for the insulin-like growth factors (IGFs) were reported in cultured breast cancer cells (Furlanetto and DiCarlo, 1984) and human breast cancer specimens (Pekonen *et al.*, 1988). IGF-I and IGF-II were found to be potent mitogens for breast cancer cells as well (Furlanetto and DiCarlo, 1984; Karey and Sirbasku, 1988). Although specific receptors for insulin and the two IGFs have been identified in breast cancer cells, there is considerable cross reactivity among these peptides and their receptors. The importance of these receptors in mediating the growth effects of insulin and the IGFs has not been clarified definitively, but current data suggest that the predominant effect of all three growth factors is mediated through the IGF-I receptor, also known as the type I somatomedin receptor (Furlanetto and DiCarlo, 1984; Coronado *et al.*, 1988).

About ten years ago we reported that epidermal growth factor (EGF) and related peptides may also modulate breast cancer cell proliferation (Osborne *et al.*, 1980; Osborne *et al.*, 1982; Arteaga *et al.*, 1988a). High affinity specific EGF receptors were identified on human breast cancer cells, and nanomolar concentrations of EGF and, subsequently, transforming growth factor-α (TGF-α) were shown to be mitogens for cultured human breast cancer cells.

More recently the intriguing observations that breast cancer cells themselves can synthesize and secrete a panoply of polypeptide growth factors or growth factor-like activities has been reported (Lippman *et al.*, 1986). Using specific cDNA probes, mRNA transcripts for TGF-α, TGF-β, IGF-I, IGF-II, pro-cathepsin D (52K protein), platelet-derived growth factor, and others have been identified in RNA extracts from breast cancer cells and tissues, and biological or immunoreactive growth factor activities have been found in conditioned medium from these cells (Coronado *et al.*, 1988; Lippman *et al.*, 1986; Peres *et al.*, 1987; Huff *et al.*, 1988; Salomon *et al.*, 1984; Nunez *et al.*, 1987; Rochefort *et al.*, 1987). Thus, breast tumors, acting like mini-endocrine glands, have the capacity to make growth factors that potentially could stimulate their own growth in an "autocrine" fashion (TGF-α or IGFs) or that could affect stromal tissues (TGF-β or PDGF) or tumor invasiveness (52K protein) by paracrine mechanisms. Of special interest is TGF-β, or family of related polypeptides that has multiple effects on a variety of tissues including stimulation of growth of mesenchymal tissues as well as growth inhibitory effects on epithelial cells including breast cancer (Arteaga *et al.*, 1988d; Knabbe *et al.*, 1987).

Even more intriguing is the observation that synthesis and secretion of several of these growth factor activities are regulated by estrogen in estrogen receptor-positive cells, while they are produced constitutively by receptor-negative cells (Lippman *et al.*, 1986). TGF-α and IGFs are increased by estrogens and decreased by antiestrogens, whereas the reverse is true with TGF-β, a potential tumor inhibitor. These observations led to the hypotheses that the growth effects of estrogens and antiestrogens may be mediated by these secreted polypeptides and that the growth advantage associated with estrogen receptor-negative tumors may be due to the constitutive production of autocrine growth factors. Recently published data do not support fully these speculations (Arteaga *et al.*, 1988a; Osborne *et al.*, 1988), but definitive conclusions await further study.

Finally, the new model of breast cancer growth regulation also proposes that stromal tissues such as fibroblasts, mononuclear cells, or endothelial cells may not be innocent bystanders in the development and growth of breast cancer. While breast cancer cells may secrete peptides that can stimulate growth of stromal elements, the "benign" stromal cells may produce diffusible products that enhance growth of the breast cancer cells (Picard *et al.*, 1986; Clemmons *et al.*, 1986). Thus, conversation among the cellular elements of a tumor may take place in the form of secreted polypeptide messengers acting via a paracrine loop.

Although the schema shown in Figure 1 adds enormous complexity to the potential breast cancer growth regulatory mechanisms, it may also offer new opportunities for clinical applications. First, the availability of recombinant natural growth inhibitors such as TGF-β offers the potential for a new treatment approach for breast cancer. Furthermore, just as recognition of the estrogen receptor biochemical pathways led to the use of receptor assays as predictive tests and to the development of effective treatments designed to block this receptor, so too might recognition of these new regulatory pathways lead to new diagnostic or treatment strategies. Finally, it is conceivable that measurements of the secreted products of breast cancer cells in body fluids or in tumor tissue itself might serve as a tumor marker/prognostic indicator for patients with this disease. In the remaining sections of this review I will focus on new experimental evidence further supporting the potential clinical applications derived from these new insights in breast cancer biology.

TGF-β, A NATURAL INHIBITOR OF BREAST CANCER GROWTH

TGF-β is a multifunctional polypeptide of 25000 daltons that has been isolated from a variety of tissues including kidney, platelets, placenta, bone and neoplastic or transformed cells (Tucker *et al.*, 1983; Massague, 1987; Sporn *et al.*, 1986). Data suggest that this peptide may play an important role in such diverse functions as bone metabolism, wound repair, or regulation of cell growth. Previous studies have reported that human breast cancer cell lines express and secrete TGF-β activity and are inhibited by exogenous TGF-β suggesting that this factor may function as an autocrine growth inhibitor (Arteaga *et al.*, 1988d; Knabbe *et al.*, 1987).

We examined the interaction of bone and platelet derived TGF-β on a panel of hormone dependent, estrogen receptor-positive and hormone independent, estrogen receptor-negative human breast cancer cells (Arteaga *et al.*, 1988d). A summary of the biologic activity of TGF-β in these cell lines using three different growth assays is shown in Table 1. Significant inhibition by TGF-β of cell growth in monolayer, cloning in soft agar, or thymidine incorporation, was observed only in the estrogen receptor-negative breast cancer cell lines. None of the receptor-positive lines was inhibited significantly by TGF-β in any of the assays including two different MCF-7 cell lines obtained from different sources.

Table 1. Biologic Activity of TGF-β in Human Breast Cancer Cell Lines

Cell Line	ER	% Inhibition Cell Number	Cloning	[^{3}H]Thymidine Incorp.
MDA-231	-	52	97	49
HS578T	-	49	ND	87
BT-20	-	25	77	ND
MDA-330	-	9	40	ND
T47D	+	0	ND	ND
ZR75-1	+	17	10	ND
CF-7L	+	13	0	0
MCF-7	+	10	0	0

Summarized from (Arteaga *et al.*, 1988d)

The explanation for the failure of the estrogen receptor-positive breast cancer cell lines to respond to exogenous TGF-β was revealed by receptor binding studies using ^{125}I-TGF-β. Typical TGF-β receptors were identified only in the four estrogen receptor-negative breast cancer cell lines. Binding was a time and temperature dependent process with higher and more rapid binding occurring at 37 C than at 0 C. However, at 37 C binding tapered off with time due to lysosomal degradation of the labeled hormone. Specificity of binding was demonstrated by the failure of EGF, insulin, IGF-I, IGF-II, and TGF-α to compete for ^{125}I-TGF-β binding. Scratchard analysis of binding data showed between 2,800 and 12,900 receptor sites per cell and a Kd ranging from 29 to 160 pM in the four receptor-negative cell lines (Table 2). Chemical cross linking studies with these cell lines revealed three different specific TGF-β binding sites with molecular weights of approximately 400,000, 92,000, and 69,000. Interestingly, none of the four estrogen receptor-positive breast cancer cell lines had detectable TGF-β binding. Because the estrogen receptor- positive and negative cell lines both secrete TGF-β activity, the possibility existed that endogenous TGF-β was occupying all or a proportion of the surface membrane receptors, thereby masking them. However, although acid-washed, estrogen receptor-negative cells demonstrated slightly higher binding of TGF-β, binding was still undetectable in the receptor-positive MCF-7 cell lines even after acid washing, suggesting that the failure to detect binding was not due to receptor occupancy by secreted TGF-β.

Table 2. Binding Profiles of TGF-β in Human Breast Cancer Cell Lines

Cell Line	Sites per Cell	Binding Affinity (Kd, pM)
MDA-231	12,900	60
HS578T	7,400	29
3T-20	2,800	130
MDA-330	5,200	160

Summarized from (Arteaga *et al.*, 1988d)

These studies suggest that there may be a relationship between the estrogen receptor status of breast cancer cells and sensitivity to TGF-β. Obviously these results employing breast cancer cell lines need confirmation using actual breast cancer tissue specimens. Our results do argue against the possibility that increased TGF-β secretion in response to tamoxifen mediates the growth inhibitory effects of the antiestrogen. Although the estrogen receptor-positive cell lines used in our studies are unresponsive to TGF-β and do not express TGF-β receptors, they have maintained their exquisite sensitivity to tamoxifen in a variety of growth assays. Finally, our results suggest that TGF-β may function as a potent growth inhibitor for certain breast cancer cells, perhaps those lacking estrogen receptor. Manipulations designed to increase endogenous secretion of TGF-β, or the administration of TGF-β *in vivo* might arrest tumor growth offering a new treatment strategy.

BLOCKADE OF GROWTH FACTOR MEMBRANE RECEPTORS, A POTENTIAL NEW TREATMENT STRATEGY

Breast cancer cells express a variety of polypeptide growth factor receptors on the surface membrane. These receptors presumably mediate the growth promoting effects of secreted factors in an autocrine fashion as well as the effects of growth factors present in blood. Just as blockade of the estrogen receptor can block estrogen's effects and inhibit growth, blockade of these membrane growth factor receptors could interrupt critical growth regulatory pathways also resulting in growth inhibition.

In preliminary studies (Arteaga *et al.*, 1988c), we have examined the effect of blockade of th type I somatomedin receptor which mediates the growth effects of insulin, IGF-I, and IGF-II. To block this receptor we utilized a mouse model monoclonal IgG_1 (α-IR_3) which blocks the binding domain of the IGF-I receptor. Similar to previous reports using other tissues or cell lines, α-IR_3 inhibited the binding of ^{125}I IGF-I in all breast cancer cell lines tested. Binding affinity of the antibody was two to five times higher than that of IGF-I in these cells.

Having demonstrated that α-IR_3 can block receptor binding, we asked whether the antibody could inhibit the growth of breast cancer cells *in vitro*. When cells were grown in medium supplemented with 10% calf serum, the antibody inhibited anchorage independent growth of six of seven breast cancer cell lines including both estrogen receptor-positive and negative. Growth inhibition was reversible by the addition of excess IGF-I suggesting that it was not due to monospecific antibody-mediated cytotoxicity. The effect of receptor blockade with α-IR_3 was next examined in cells growing under serum-free conditions. If secretion of IGFs by the breast cancer cells is an important autocrine growth mechanism, then blockade of the type I somatomedin receptor should inhibit basal cell proliferation in the absence of serum. The antibody blocked the stimulatory effects of IGF-I and IGF-II in breast cancer cells growing in serum-free medium. However, basal growth in serum-free medium was not inhibited by α-IR$_3$ when either ^{3}H-thymidine incorporation or growth in monolayer was assessed. In the MCF-7 line stimulation of DNA synthesis and cell proliferation was observed with α-IR$_3$, suggesting that it may even function under these conditions as a partial agonist.

Since α-IR$_3$ was able to antagonize the effects of exogenous IGF-I, we hypothesized that it might also antagonize the effect of secreted IGF activity by the cells in response to estrogen, thereby inhibiting or blocking estrogen-induced growth. However, the antibody failed to inhibit estrogen-induced increases in thymidine incorporation or monolayer growth suggesting that the IGF activity secreted into the culture medium in response to estrogen is not a primary mediator of estrogen-induced growth.

To determine whether blockade of the type I somatomedin receptor might inhibit breast cancer *in vivo*, athymic nude mice were inoculated with either hormone responsive MCF-7 breast cancer cells or estrogen receptor-negative MDA-231 cells (Arteaga *et al.*, 1988c). Mice were simultaneously begun on varying doses of α-IR$_3$ administered IP twice weekly for four to five weeks. *In vivo* growth of the MDA-231 cell was markedly inhibited by α-IR$_3$ in a dose response fashion. Tumor "take rate" was reduced as was tumor size in those tumors that did develop. Inhibition of growth was not observed using a control antibody of the same class. Histologic examination of tumors from treated mice revealed marked fibrosis and tumor cell loss. In contrast to the effects of *in vivo* treatment with α-IR$_3$ on MDA-231 tumor growth, no effect of antibody therapy was observed with mice inoculated with MCF-7 tumors. The failure of inhibition of MCF-7 tumor growth by blockade of the IGF-I receptor suggests that other growth regulatory pathways may be more important or may be able to circumvent blockade of this single pathway in these cells. Nevertheless, these data demonstrate for the first time that blockade of a polypeptide growth factor pathway *in vivo* can alter growth of human breast cancer cells. Other reports have shown that blockade of the EGF receptor or antibody neutralization of bombesin (Masui *et al.*, 1984; Cuttitta *et al.*, 1985) can inhibit growth of other tumors. These data support the hypothesis that antagonism of these pathways may offer a new strategy for "endocrine" therapy of breast cancer.

AUTOCRINE GROWTH FACTORS: POTENTIAL TUMOR MARKERS

Secreted products from tumors have proven extremely valuable as tumor markers. Assays for such products as β-human chorionic gonadotropin, α-feto protein or carcinoembryonic antigen have proven useful in the detection and evaluation of several malignant diseases. If secreted products from breast cancer cells enter the blood or other body fluids in sufficient concentrations, then their measurement might also prove useful as a tumor marker. Measurement of these factors could provide clues for early diagnosis, and their quantification might theoretically have prognostic significance. Those tumors secreting higher levels of growth factors might be expected to have a growth advantage and thus a worse prognosis.

Since human breast cancer cells have been shown to express and secrete TGF-α, we have been interested in developing assays to detect this polypeptide in body fluids. Initially, Hanauske *et al.* (1988) used a radioimmunoassay for TGF-α to demonstrate that levels of this growth

Table 3. TGF-α Immunoreactivity in Effusions from Cancer Patients

Tumor Type	(n)	Detectable TGF-α
Breast	(34)	38%
Ovary	(31)	43%
Lung	(24)	50%
Unknown primary	(12)	50%
Colon	(8)	50%
Others	(21)	33%

factor in malignant effusions from cancer patients correlated with the ability of tumor cells from those effusions to form colonies in soft agar. With this data suggesting that TGF-α activity in effusions from cancer patients correlated with growth in soft agar, one parameter of tumorigenicity, we hypothesized that it might also correlate with clinical variables usually predictive of poor prognosis.

TGF-α activity was measured in 130 effusions from patients with various types of cancer with a radioimmunoassay using sheep antibodies against the C-terminal 17 amino acids of linear rat TGF-α (Arteaga *et al.*, 1988b). Seventeen effusions from non-cancer patients served as controls. Only 3 of 17 control effusions had detectable TGF-α activity and all of these were less than 2 ng/ml. None of the five exudative effusions in control patients had detectable TGF-α. In contrast, 42% of the effusions from cancer patients contained immunoreactive TGF-α activity (Table 3). Concentrations ranged from 1.5 to 50 ng/ml. Interestingly, TGF-α was detected in effusions from cancer patients regardless of whether the effusions were cytology positive or negative for malignant cells. TGF-α was detected in 40 of 89 (45%) of cytology positive effusions and 15 of 41 (37%) of cytology negative specimens. This suggests that the TGF-α activity in the effusion does not depend on the actual presence of malignant cells in the effusion and that it may originate from distant metastatic sites via the systemic circulation.

The presence of immunoreactive TGF-α activity in the effusions correlated with several clinical variables. When all tumor types were considered, or when just the subset of patients with breast cancer was examined, TGF-α activity correlated with the patients' performance status and tumor burden estimated by the number of sites of disease. TGF-α was detected in only 27% of effusions from patients with a relatively good performance status of ≥ 2, but it was present in 70% of those with a performance status of 3 or more. Similarly, only 4% of patients with a single site of metastatic disease had detectable TGF-α in the pleural effusion, while 48% of those with two sites, and 97% of those with > 2 sites of metastasis had detectable levels.

Similar correlations were found when the subset of patients with breast cancer was considered (Table 4). Immunoreactive TGF-α was detectable in a significantly greater percentage of breast cancer patients with a poor performance status, > 2 metastatic sites of disease, premenopausal, or those with estrogen and progestin receptor-negative tumors. Each of these factors is generally considered to be an adverse prognostic sign for patients with metastatic breast cancer. In view of the *in vitro* observations using cultured breast cancer cells demonstrating that estrogen receptor-negative cell lines secrete high levels of TGF-α constitutively, it is interesting that TGF-α activity is more

Table 4. Correlation of TGF-α Activity in Breast Cancer Patients With Clinical Variables

Clinical Variable	Detectable TGF-α
Performance Status	
≥ 2	11%
> 2	60%
Metastatic Sites	
≥ 2	7%
> 2	100%
Menopausal Status	
Pre	61%
Post	25%
ER Status	
+	9%
-	56%
PgR Status	
+	6%
-	77%

likely to be present in effusions from patients with receptor-negative tumors.

The survival of patients in this study also correlated with the presence or absence of immunoreactive TGF-α in the effusions. The median survival for the 72 patients with no detectable TGF-α activity was twice as long (6 months) as that for the 48 patients with detectable TGF-α (3 months. These survival differences were still observed when only the good prognosis patients with a good performance status were considered. The breast cancer patients with TGF-α activity in their effusions also showed a statistically significant shorter survival compared to those without TGF-α activity. A Cox model multivariate analysis which included TGF-α activity, performance status, and number of sites of disease, showed that TGF-α activity and performance status retained a significant inverse relationship with survival. These results suggest that, although related to performance status and number of sites of disease, the presence of TGF-α activity in effusions from this group of patients with advanced cancer provides additional independent prognostic information.

These data demonstrate that TGF-α can be measured in a body fluid in patients with advanced cancer and that the activity correlates with important clinical variables. Obviously, measurement of a tumor marker in pleural or ascitic fluid is not practical and has little clinical utility. Nevertheless, these results are encouraging and suggest that further studies of the measurement of TGF-α as well as other secreted growth factors in more accessible body fluids such as urine or serum, are necessary to fully evaluate these potential tumor markers.

To summarize, proliferation of human breast cancer cells is regulated by a complex interaction among several steroid and polypeptide hormones and growth factors. Furthermore, breast cancer cells can make and secrete growth factors that may have important autocrine growth effects on the tumor, paracrine effects on stromal tissues, or endocrine effects on the host. Additional studies are needed to determine if expression of these growth factors by the tumor has prognostic significance, or whether antagonism of these biochemical pathways may offer a new approach to treatment.

REFERENCES

Arteaga C.L., Coronado E., and Osborne C.K. 1988a. Blockade of the epidermal growth factor receptor inhibits transforming growth factor α-induced but not estrogen-induced growth of hormone-dependent human breast cancer. Mol. Endocrinol. (in press).

Arteaga C.L., Hanauske A.-R., Clark G.M., Osborne C.K., Hazarika P., Pardue R.L., and Von Hoff D.D. 1988b. Immunoreactive α transforming growth factor (IraTGF) activity in effusions from cancer patients: A marker of tumor burden and patient prognosis. Cancer Research. (in press).

Arteaga C.L., Kitten L., Coronado E., Jacobs S., Kull F., and Osborne C.K. 1988c. Blockade of the type I somatomedin receptor inhibits growth of estrogen receptor negative human breast cancer cells in athymic mice. Proc. 70th Annual Meeting Endocrine Society 191:683 (Abstract).

Arteaga C.L., Tandon A.K., Von Hoff D.D., and Osborne C.K. 1988d. Transforming growth factor β: Potential autocrine growth inhibitor of estrogen receptor-negative human breast cancer cells. Cancer Research 48:3898.

Clemmons D.R., Elgin R.G., Han V.K.M., Casella S.V., D'Ercole A.J., and Van Wyk J.J. 1986. Cultured fibroblast monolayers secrete a protein that alters the cellular binding of somatomedin-C/insulin like growth factor I. J. Clin. Invest. 77:1548.

Coronado E., Ramasharma K., Li C.H., Kitten L., Marshall M., Fuqua S., and Osborne K. 1988. Insulin-like growth factor II (IGF-II): A potential autocrine growth factor for human breast cancer. Proc. Am. Assoc. Cancer Res. 29:237 (Abstract).

Cuttitta F., Carney D.N. Mulshine J., Moody T.W., Fedorko J., Fischler A., and Minna J.D. 1985. Bombesin-like peptides can function as autocrine growth factors in human small-cell lung cancer. Nature (London) 309: 823.

Furlanetto R.W. and DiCarlo J.N. 1984. Somatomedin-C receptors and growth effects in human breast cells maintained in long-term tissue culture. Cancer Research 44:2122.

Hanauske A.R., Arteaga C.L., Clark G.M., Buchok J., Marshall M., Hazarika P., Pardue R.L., and Von Hoff D.D. 1988. Determination of transforming growth factor activity in effusions from cancer patients. Cancer (Phil.) 61:1832.

Huff K.K., Knabbe C., Lindsey R., Kaufman D., Bronzert D., Lippman M.E., and Dickson R.B. 1988. Multihormonal regulation of insulin-like growth factor-I-related protein in MCF-7 human breast cancer cells. Mol. Endocrinol. 2:200.

Ikeda T., Danielpour D., and Sirbasku B.A. 1983. Isolation and properties of endocrine and autocrine type mammary tumor cell growth factors (estromedins), In: "Progress in Cancer Research and Therapy", Vol. 31 (F. Bresciani *et al.*, eds.) Raven Press. New York. p.144.

Karey K.P. and Sirbasku D.A. 1988. Differential responsiveness of human breast cancer cell lines MCF-7 and T47D to growth factors and 17-β-estradiol. Cancer Research 48:4083.

Knabbe C., Lippman M.E., Wakefield L.M., Flanders K.C., Kasid A., Derynck R., and Dickson R.B. 1987. Evidence that transforming growth factor-β is a hormonally regulated negative growth factor in human breast cancer cells. Cell 48:417.

Lippman M.E., Dickson R.B., Bates S., Knabbe C., Huff K., Swain S., McManaway M., Bronzert D., Kasid A., and Gelmann E.P. 1986. Autocrine and paracrine growth regulation of human breast cancer. Breast Cancer Res. Treat. 7:59.

Lippman M., Monaco M.E., and Bolan G. 1977. Effects of estrone, estradiol, and estriol on hormone-responsive human breast cancer in long-term tissue culture. Cancer Research 37:1901.

Massague J. 1987. The TGF-β family of growth and differentiation factors. Cell 49:437.

Masui H., Kawamoto T., Sato J.D., Wolf B., Sato G., and Mendelsohn J. 1984. Growth inhibition of human tumor cells in athymic mice by anti-epidermal growth factor receptor monoclonal antibodies. Cancer Research 44:1002.

Nunez A.-M., Jakowlev S., Briand J.-P., Gaire M., Krust A., Rio M.-C., and Chambon P. 1987 Characterization of the estrogen-induced pS2 protein secreted by the human breast cancer cell line MCF-7. Endocrinology 121:1759.

Osborne C.K., Bolan G., Monaco M.E., and Lippman M.E. 1976. Hormone responsive human breast cancer in long term tissue culture: Effect of insulin. Proc. Natl. Acad. Sci. (USA) 73:4536.

Osborne C.K., Hamilton B., and Nover M. 1982. Receptor binding and processing of epidermal growth factor by human breast cancer cells. J. Clin. Endocrinol. Metab. 55:86.

Osborne C.K., Hamilton B., Titus G., and Livingston R.B. 1980. Epidermal growth factor stimulation of human breast cancer cells in tissue culture. Cancer Research 40:2361.

Osborne C.K., Monaco M.E., Lippman M.E., and Kahn C.R. 1978. Correlation among insulin binding, degradation, and biological activity in human breast cancer cells in long-term tissue culture. Cancer Research 38:94.

Osborne C.K., Ross C.R., C.R., Coronado B.S., Fuqua S.A.W., and Kitten, L. J. 1988. Secreted growth factors from estrogen receptor-negative human breast cancer do not support growth of estrogen receptor-positive breast cancer in the nude mouse model. Breast Cancer Res. Treat. 11:211.

Peokonen F., Partanen S., Makinen T., and Rutanen E.-M. 1988. Receptors for epidermal growth factor and insulin-like growth factor I and their relation to steroid receptors in human breast cancer. Cancer Research 48:1343.

Peres R., Betsholtz C., Westermark B., and Heldin C.-H. 1987. Frequent expression of growth factors for mesenchymal cells in human mammary carcinoma cell lines. Cancer Research 47:3425.

Picard O., Rolland Y., and Poupon M.F. 1986. Fibroblast-dependent tumor-igenicity of cells in nude mice: Implication for implantation of metastases. Cancer Research 46:3290.

Rochefort H., Capony F., Garcia M., Vacailles V., Freiss G., Chambon M., Morisset M., and Vignon F. 1987. Estrogen-induced lysosomal proteases secreted by breast cancer cells: a role in carcinogenesis? J. Cell Biochem. 35:17.

Salomon D.S., Zwiebel J.A., Bano M., Losonczy I., Fehnel P., and Kidwell W.R. 1984. Presence of transforming growth factors in human breast cancer cells. Cancer Research 44:4069.

Sporn M.B., Roberts A.B., Wakefield L.M., and Assoian R.K. 1986. Transforming growth factor-β: Biological function and chemical structure. Science 233:532.

Tucker R. F., Volkenaut M.E., Branum E.L., and Moses E.L. 1983. Comparison of intra- and extracellular transforming growth factors from nontransformed and chemically transformed mouse embryo cells. Cancer Research 43:1581.

Welsch C.W. and Nagasawa H. 1977. Prolactin and murine mammary tumorigenesis: A review. Cancer Research 37:951.

THE ROLES OF TRANSFORMING GROWTH FACTORS α AND β IN GROWTH REGULATION OF NORMAL AND TRANSFORMED HUMAN MAMMARY EPITHELIAL CELLS

Eva M. Valverius
Marc E. Lippman and
Robert B. Dickson

Vincent Lombardi Cancer Center
Georgetown University Hospital
3900 Reservoir Road N.W.
Washington, D.C. 20007

AUTOCRINE GROWTH FACTORS IN HUMAN BREAST CANCER

Abnormal expression of specific growth factors or their receptors may be involved in the initiation and development of a variety of malignancies (Heldin and Westermark, 1984; Goustin *et al.*, 1986). Normal or nontransformed cells in culture generally exhibit a high degree of dependence upon exogenously supplied growth factors for proliferation, whereas transformed cells demonstrate a partial or complete relaxation in their growth factor requirements. This independence from exogenous growth factors may result from an increased level of expression of the same growth factors and/or receptor "down regulation" in cells which have become malignantly transformed (Sporn and Roberts, 1985). It has been postulated that by an autocrine feed-back loop, the secreted growth factors could act on the cells' own receptors and thus contribute to an abnormal growth pattern.

The mechanisms regulating growth of malignant mammary cell growth are poorly understood. Much circumstantial evidence indicates a role for estrogen in enhancing the expression of certain protooncogenes and/or growth factors. In the case of hormone responsive breast cancer cells, growth stimulation by estrogen is accompanied by an increase in transforming growth factor α (TGFα) production (Bates *et al.*, 1988a), whereas growth inhibition of a breast cancer cell line by an antiestrogen is paralleled by augmented secretion of transforming growth factor β (TGFβ) (Knabbe *et al.*, 1987). With hormone-independent breast cancer cell lines, compared to hormone-dependent breast cancer cells, the production of both of these growth factors, as well as many other growth regulatory peptides, is elevated (Bates *et al.*, 1986; Artega *et al.*, 1988; Dickson and Lippman, 1988), thus implying a role for growth factors in the expression of a more malignant phenotype and escape from normal hormonal control.

Transforming growth factors derive their name from their ability to reversibly induce the transformed phenotype (specifically defined as the capacity for anchorage-independent growth) in certain rodent fibroblasts

(Todaro *et al.*, 1983). They are polypeptides initially found to be secreted by a variety of retrovirally, chemically, or oncogene-transformed human and rodent cell lines (Sporn and Roberts, 1985). The two major classes of structurally and functionally distinct transforming growth factors are TGFα and TGFβ. TGFα and TGFα-like peptides exist as multiple species ranging from apparent sizes 6 to 35 kDa (Bates *et al.*, 1986) and compete with the structural homolog epidermal growth factor for binding to the same receptor (Massague, 1983). TGFβ consists of at least two different gene products, each forming 25 kDa dimeric species (Cheifetz *et al.*, 1988). There is a complex pattern of interactions of these species with the TGFβ receptor, which has been described as three different molecular weight species. TGFβ, and more recently TGFα, have been found in urine and pleural and peritoneal effusions of cancer patients (Stromberg *et al.*, 1987; Artega *et al.*, 1988; Sairenji *et al.*, 1987). They have also been found in some normal tissues (Derynck, 1988; Sporn and Roberts, 1986). It has been proposed that transformation of cells from normal to malignant growth patterns may involve either increased production of growth stimulatory factors or decreased production of growth inhibitory substances, or altered responsiveness to either or both of these groups of growth factors (Sporn and Roberts, 1985).

Essential to understanding the pathways of growth control in human neoplastic cells is a knowledge of normal growth regulation. To date, this area of investigation has lagged behind in studies of the influence of growth factors and their relationships to transformation by oncogenes in epithelial cell systems due to the difficulties involved in culture of normal epithelial cells. However, the recent development of specialized serum-free culture conditions (Stampfer, 1985; Hammond *et al.*, 1984) has facilitated the study and definition of regulatory pathways and growth factors in normal human keratinocytes (Coffey *et al.*, 1987), normal human bronchial epithelial cells (Masui *et al.*, 1986), and, as will be discussed in this paper, normal human mammary epithelial cells.

EGF AND TGFα *IN VIVO* AND *IN VITRO*

Epidermal growth factor (EGF) regulates the proliferation and differentiation of the mouse mammary gland *in vivo* and mouse mammary explants *in vitro* (VonderHaar, 1988; Oka, 1988). EGF is also a required supplement for the *in vitro* growth of normal human mammary epithelial cells (Stampfer, 1985). Human breast cancer cells, however, no longer require exogenous EGF for continuous growth, although many breast cancer cell lines retain growth stimulatory responses to EGF (Osborne *et al.*, 1980, Davidson *et al.*, 1987). TGFα, a functional homolog of EGF, can produce essentially the same biological effects in mouse mammary explants and cultured human and mouse mammary epithelial cell lines as EGF (VonderHaar, 1988; Salomon *et al.*, 1987), but its role in normal or malignant mammary development is not yet clear. Mouse salivary gland derived EGF appears to be necessary for spontaneous mammary tumor formation in the mouse model (Kurachi *et al.*, 1985) as well as for growth of the tumors once they are formed. EGF can also partially replace estrogen to promote tumorigenicity of a human breast cancer cell line in nude mice (Dickson *et al.*, 1986).

TGFα has been directly implicated in cellular transformation in a number of studies. For example, overexpression of TGFα following transfection of a human TGFα cDNA expression vector into the immortal non-tumorigenic mouse mammary epithelial cell line NOG-8 led to tumorigenicity and capacity for anchorage-independent growth (Shankar *et al.*,

1989). Likewise, in two of three studies using rodent fibroblasts as recipients for human or rat TGFα cDNA, transformation was also achieved (Rosenthal *et al.*, 1986; Watanabe *et al.*, 1987), whereas in the third study, TGFα induced proliferation but not malignant progression (Finzi *et al.*, 1987). EGF can also act as a transformation inducing agent when transfected and overexpressed in rodent fibroblasts (Stern *et al.*, 1987). Furthermore, a direct correlation between the degree of TGFα production, ras expression and the degree of malignant transformation has been demonstrated in a recent study utilizing a glucorticoid-inducible point-mutated c-Ha-ras construct transfected into immortal mouse mammary epithelial cells (Ciardiello *et al.*, 1988). In studies of human pathological specimens, TGFα mRNA was detected in 70% of primary human breast cancers (Bates *et al.*, 1988a) and in approximately 30% of benign breast lesions (Travers *et al.*, 1988). Immunoreactive TGFα has been found in fibroadenomas and 25-50% of primary human mammary carcinomas (Macias *et al.*, 1987; Perroteau *et al.*, 1986).

RAS IN BREAST CANCER

In the classical studies of rodent fibroblasts transformed by Rous sarcoma virus, v-Ha-ras, or v-Ki-ras, (DeLarco and Todaro, 1978; Anzano *et al.*, 1983), increased production of "sarcoma growth factor", which later would come to be characterized as consisting of both TGFα and TGFβ, was demonstrated. Similarly, increased production of TGFα has been reported following transfection of MCF-7 human breast cancer cells with v-Ha-ras or of mouse mammary epithelial cells by a point-mutated human c-Ha-ras gene (Dickson *et al.*, 1987; Salomon *et al.*, 1987). Whereas point-mutations of the c-Ha-ras and c-Ki-ras proto-oncogenes have been observed in two hormone-independent, human breast cancer cell lines (Kraus *et al.*, 1984; Kozma *et al.*, 1988), the role of ras expression in clinical cases of human breast cancer has not yet been clarified. One study indicates a positive correlation between expression of p21 ras protein and progression of human breast cancer (Clair *et al.*, 1987). In another study, no such correlation could be observed, although malignant and dysplastic breast lesions did have elevated levels of p21 ras protein compared to normal tissues (Horan-Hand *et al.*, 1987). Neither of these studies addressed the state of activation (or mutation) of the ras gene, but studies by several other groups indicate that ras activation along with some additional event(s) are necessary for neoplastic transformation (Medina, 1988).

THE EGF/TGFα RECEPTOR

The role of TGFα or EGF in transformation may also involve alterations in the function of their receptor, the EGF receptor. Clinical evidence for the role of increased expression of the EGF receptor, or its structurally related homolog c-erbB-2, in more aggressive and hormone unresponsive breast cancer has accumulated in recent years (Sainsbury *et al.*, 1987; Perez *et al.*, 1984; Slamon *et al.*, 1987). This is also supported by *in vitro* studies of cultured human breast biopsies (Spitzer *et al.*, 1987) and in established human breast cancer cell lines (Davidson *et al.*, 1987). In transfection studies on rodent fibroblasts, overexpression of the EGF receptor can induce the transformed phenotype upon stimulation by EGF (Velu *et al.*, 1987; Di Fiore *et al.*, 1987a; Riedel *et al.*, 1988). Likewise, transfection of rodent fibroblasts with c-erbB-2, structurally related to the EGF receptor but lacking EGF binding capacity, results in transformation (Hudziak *et al.*, 1987; Di Fiore *et al.*, 1987b).

TGFβ *IN VIVO* AND *IN VITRO*

TGFβ is a bifunctional regulator of cell growth (Moses *et al.*, 1985; Roberts *et al.*, 1985). Depending on the presence of serum or other growth factors, it can either stimulate or inhibit cellular proliferation. Generally, cells of mesenchymal origin, such as fibroblasts, respond to TGFβ by growth stimulation, whereas most cells of epithelial origin are inhibited by TGFβ to varying degrees. TGFβ also has a multitude of other effects on various cell systems, such as induction of differentiation of bronchial epithelial cells and chondrocytes (Masui *et al.*, 1986; Seyedin *et al.*, 1986) and stimulation of synthesis of extracellular basement membrane (Roberts *et al.*, 1986; Ignotz and Massague, 1986). In fibroblasts, it has been proposed that TGFβ exerts its growth promoting effects via c-<u>sis</u> induction (Leof *et al.*, 1986). In studies of rodent endothelial cells and in a breast cancer cell line with over-expression of the EGF receptor (Takehara *et al.*, 1987; Fernandez-Pol *et al.*, 1987), effects of TGFβ on growth were linked to alterations in EGF receptor binding characteristics.

TGFβ mRNA has been found in a majority of human tissues studied to date, unlike the fairly restricted distribution of TGFα. The relative abundance of TGFβ in platelets (Assoian *et al.*, 1983) and bone (Seyedin *et al.*, 1986) as well as its stimulatory action on the production of structural and extracellular matrix components, indicate that TGFβ may have a role in the processes of wound repair and tumor cell metastasis (Sporn and Roberts, 1986). Expression of TGFβ mRNA has been shown to be more abundant in malignant human breast biopsies than in benign lesions (Travers *et al.*, 1988) and in phenotypically more aggressive hormone independent breast cancer cell lines compared to hormone dependent breast cancer cells (Bates *et al*, 1986; Artega *et al.*, 1988). In the developing normal mouse mammary gland, exogenous TGFβ was shown to be a potent, reversible *in vivo* inhibitor of terminal end bud formation (Silberstein and Daniel, 1987). In most *in vitro* systems, however, no obvious correlation has been observed between expression of the transformed phenotype and resistance to growth inhibition by TGFβ, largely because of the lack of complete studies comparing transformed cells to their normal parental cells. In one study, rodent fibroblasts transformed by Harvey or Moloney sarcoma viruses demonstrated increased production of TGFβ and reduced TGFβ binding (Anzano *et al.*, 1987), but the growth responses to TGFβ were not examined. Another study using rodent embryo fibroblasts transformed by the activated c-Ha-<u>ras</u> gene showed reduced responsiveness to TGFβ (Leof *et al*, 1987), but no information on expression of TGFβ or its receptor was given. Immortalized human bronchial epithelial cells when transformed by activated v-Ki-<u>ras</u> showed loss of responsiveness to TGFβ compared to their untransformed counterparts (Reddel *et al*, 1989), but again, expression of TGFβ or its receptor are not described. In two studies of TGFβ responsiveness comparing established cancer cell lines to normal cell lines derived from the same organ system, an altered responsiveness to TGFβ is seen in the neoplastic cell lines with the loss of responsiveness correlated to an absence of receptors for TGFβ (squamous cell carcinoma cell lines compared to hormal human prokeratinocytes, Shipley *et al.*, 1986; retinoblastoma-derived cell lines compared to human fetal retinal cells, Kimchi *et al.*, 1988). Since the cell types in these studies may not have been derived from the same original stem cell population, however, it is difficult to draw any conclusions regarding the role of loss of TGFβ receptors in malignant progression.

TGFα AND EGF RECEPTOR EXPRESSION IN HUMAN MAMMARY EPITHELIAL CELLS

In our studies, we have examined the growth responsiveness of normal, immortalized, and oncogene-transformed human mammary epithelial cells to EGF, TGFα and TGFβ. We wanted to determine whether expression of the malignant phenotype could be correlated with alterations in the responsiveness to EGF/TGFα or with escape from growth inhibition by TGFβ, and whether any such alterations would be accompanied by modulation of EGF or TGFβ cell surface receptor characteristics. Transformation may involve either an increased positive autocrine stimulation via increased TGFα production, or a reduction in negative autocrine regulation by decreased TGFβ production. We therefore also sought to assess whether there would be any quantitative or qualitative differences between the normal and neoplastic cells in their possible production of TGFα or TGFβ.

We have utilized a series of human mammary epithelial cells derived from histopathologically normal breast tissue obtained from reduction mammoplasty on young, non-pregnant, non-lactating women (Stampfer and Bartley, 1988). The 184, 172, and 161 cells are normal diploid epithelial cell strains capable of growing for 12-20 passages (50-70 cell doublings) on plastic surfaces before undergoing senescence (or terminal differentiation). They require a MCDB 170-based medium containing bovine pituitary extract, EGF, insulin, and hydrocortisone, among other supplements, for rapid proliferation and serial passage. The epithelial origin of these cells has been established using immunocytochemical markers and electron microscopy (Stampfer and Bartley, 1988). Following benzo-a-pyrene treatment of the 184 cells, an immortalized cell line was established and cloned, 184A1N4 (Stampfer and Bartley, 1985), and used as recipient for various oncogenes carried in retroviral vectors. The resulting (not clonally selected) cell lines carry and express v-Ha-ras (184A1N4-H), v-mos (184A1N4-M), SV40 T (184A1N4-T), or both v-Ha-ras and SV40 T (184A1N4-TH), or both v-Ha-ras and v-mos (184A1N4-MH) (Clark *et al.*, 1988).

The immortalized 184A1N4 cells can be propagated in a less complex medium than the parent 184 cells, utilizing IMEM with 0.5% fetal calf serum (FCS), EGF, insulin and hydrocortisone. Like the 184 cells, they will not grow under anchorage-independent conditions nor form tumors in nude mice. All the oncogene transformed cells are grown in IMEM with 10% FCS. The 184A1N4-T and 184A1N4-M cells clone poorly in soft agar and are not tumorigenic. The 184A1N4-H and 184A1N4-MH cells are weakly or moderately tumorigenic, respectively, but neither cell line clones well in soft agar. Phenotypically, the 184A1N4-TH cells are fully transformed in that they grow extensively under anchorage-independent conditions and are highly tumorigenic in nude mice. Neither the normal parental 184 cells nor any of the immortalized oncogene transformed sublines express the estrogen receptor (Valverius, unpublished data).

Since initially the most obvious difference in growth requirements among the 184-derived cells was the capability of the transformed cells to grow in a simple, nondefined medium, we first decided to study the cells' responsiveness to EGF or TGFα (Valverius *et al.*, 1989). In assays for growth response to EGF/TGFα, we found that the degree of responsiveness diminished with increased expression of the transformed phenotype. The normal parental 184 cells and the immortalized 184A1N4 cells were dependent on EGF at clonal densities. In mass culture, however, the 184 cells grew quite well without EGF during the time of the assays (which did not involve passaging). As will be explained later, we have attributed this density-related EGF-dependence in the 184 cells to crossfeeding and autocrine stimulation of growth at higher cell densities. Among the oncogene carrying cells, all cell lines expressing v-Ha-ras showed lack of responsiveness to exogenous EGF or TGFα under both anchorage-dependent and anchorage-independent conditions. Growth of 184A1N4-M and 184A1N4-T, on the other hand, were both stimulated by

EGF/TGFα. Also, while normally not capable of significant anchorage-independent growth, the 184A1N4-T could be induced to clone in soft agar with up to 7% cloning efficiency by supplementation with either EGF or TGFα.

To evaluate whether these marked differences in cellular responsiveness to exogenous EGF/TGFα related to differences in production of endogenous TGFα, we first measured TGFα mRNA expression by Northern analysis of total cellular RNA. We found no differences in abundance of the expected 4.8 kb TGFα species among all 184-derived cells. In addition, all of the cells had TGFα mRNA levels comparable to those of the hormone-independent human breast cancer cell lines MDA-MB-231 and MDA-MB-468. Five other normal, non-immortal human breast epithelial cell strains also showed the same high TGFα expression (Bates *et al.*, 1988b). We next determined the amounts of TGFα-like activity secreted into medium which had been conditioned by subconfluent cells during a 48-hour period. As assayed either by a TGFα-specific radioimmunoassay (Linsley *et al.*, 1985) or for induction of anchorage-independent cloning of normal rat kidney cells in soft agar (Bates *et al.*, 1986), all oncogene expressing cells produced similar levels of bioactive and immunoreactive TGFα. In fact, the levels were comparable to those found in the MDA-MB-231 cell line. The 184 cells, however, when grown in the complete medium containing EGF and bovine pituitary extract, produced 5 to 10-fold more immunoreactive TGFα than all the other cell lines. Even after withdrawal of EGF, bovine pituitary extract, and insulin from the medium prior to conditioning, the 184 cells produced similar amounts of immunoreactive TGFα and almost three-fold more bioactive TGFα, compared to the rest of the 184-derived cell lines. This might explain the less stringent growth requirements for EGF supplementation observed for the 184 cells at high cell densities. The 184 cells in high density cultures could presumably produce sufficient amounts of TGFα to stimulate their own proliferation through an autocrine mechanism, thereby rendering them relatively independent of exogenous EGF or TGFα.

The most important difference between the normal human mammary epithelial cells used in this study and normal tissue, is that these cells are rapidly proliferating in culture. It could thus be postulated that TGFα production is more closely associated with cellular proliferation rather than serving as a marker for transformation, at least in this system. To test this hypothesis, we studied TGFα mRNA expression by Northern analysis and by *in situ* hybridization techniques in the nonproliferating organoid preparations from which cultures of normal mammary epithelial cell strains are initiated. We found that TGFα mRNA was either not detectable or expressed at very low levels in the organoids (Bates *et al.*, 1988b). Likewise, TGFα expression declined in low density cultures of the 184 cells upon withdrawal of EGF and bovine pituitary extract from the medium. Thus, we conclude that, in human mammary epithelial cells, TGFα expression may be an indicator for conversion to a proliferative state and is not directly associated with oncogene mediated transformation.

Since the observed phenotypic differences among the 184-derived cells could not be explained on the basis of variations in TGFα production and since variations in growth factor responsiveness and a possible autocrine loop may occur at the receptor level, we next examined EGF receptor expression in the human mammary epithelial cells. We could find no significant differences in the levels of expression of two EGF receptor specific mRNA species, 10 and 5.6 kb, among the normal and oncogene transformed cells using Northern analysis of total cellular RNA. Southern analysis of HindIII digested DNA also failed to show any differences since no apparent EGF receptor gene amplifications or rearrangements were observed. EGF receptor binding characteristics were subsequently determined on all the human mammary epithelial cell lines. The 184, 184A1N4 and 184A1N4-M cells all had 3-4x10^5 binding sites per cell. In the 184A1N4 and 184A1N4-M cells both a high and low affinity, EGF-binding component were determined, while only the high affinity component was evident in the 184 cells. The 184A1N4-T cells had markedly elevated levels of both binding components, with a total number of binding sites per cell close to 2x10^6. This was especially notable since these cells could respond with induction of anchorage-independent growth upon EGF stimulation. Other studies have demonstrated that overexpression of certain oncogenes can in some cells elicit hypersensitivity to EGF. This response, however, usually occurred without concomitant changes in either receptor binding or cellular phenotypic characteristics (e.g. c-myc overexpression in rat fibroblasts, Stern *et al.*, 1986; c-myc overexpression in chicken mesenchymal cells, Balk *et al.*, 1985; pp$60^{c\text{-}src}$ overexpression in rat embryo fibroblasts, Luttrell *et al.*, 1988). In addition, increased number of EGF receptor sites/cell in the 184A1N4-T cells was associated with growth stimulation. This contrasts with the response of two other cell lines to EGF where EGF receptor overexpression also has been observed (the human breast cancer cell line MDA-MB-468, Filmus *et al.*, 1985; the epidermoid cell line A431, Santon *et al.*, 1986). In these cells, EGF causes growth inhibition.

All three 184-derived cell lines expressing v-Ha-ras were found to have a slight reduction in number of EGF binding sites per cell, 1.7-3x10^5, and lacked the high affinity EGF binding component. These findings are in agreement with several other studies which have demonstrated a reduction in high affinity EGF receptor binding in the presence of v-Ha-ras or activated c-Ha-ras (Salomon *et al.*, 1987; Ciardiello *et al.*, 1988; Kamata and Feramisco, 1984). Our findings differ from the total absence of EGF binding originally observed in virus-transformed rodent fibroblasts, however (Todaro *et al.*, 1976). It is not yet clear as to what the exact role the loss of the high affinity binding component of the EGF receptor may be or how it may influence cellular responsiveness to EGF. Nevertheless, it is unlikely that the relatively slight reduction in total EGF binding sites is of major importance since there are a number of human breast cancer cell lines which exhibit only a fraction of the number of total EGF binding sites that was observed in the 184-derived cells, yet demonstrate growth sensitivity to EGF (Osborne *et al.*, 1980).

TGFβ AND HUMAN MAMMARY EPITHELIAL CELLS

As discussed above, the transition from normal to malignant growth patterns may also involve escape from normal growth inhibitory mechanisms. We therefore decided to study the response of normal, immortalized and oncogene transformed human mammary epithelial cells to TGFβ. We have earlier presented evidence that TGFβ is a hormonally regulated negative growth factor for a hormone responsive human breast cancer cell line (Knabbe *et al.*, 1987). Others have shown that TGFβ is a potential autocrine growth inhibitor for four hormone independent human breast cancer cell lines (Artega *et al.*, 1988). Also, resistance to the growth inhibitory effects of TGFβ is induced in rat liver epithelial cells by transfection with an activated v-Ha-ras oncogene (Houck *et al.*,

1987). We found that the normal human breast epithelial cell strain 184 was markedly growth inhibited by TGFβ, while the immortalized subclone 184A1N4 was much less sensitive (Valverius *et al.*, 1988). Under anchorage-dependent conditions, the 184A1N4-TH were the least sensitive of all the cells, and in soft agar assays, the 184A1N4-TH were not at all inhibited by TGFβ. The proportion of active to latent TGFβ produced by the cells increased slightly with oncogene transformation, but the amount of total TGFβ produced as well as the level of TGFβ mRNA, remained essentially unaffected. Likewise, TGFβ receptor binding parameters did not change between the 184A1N4 cells and the oncogene transformants. Thus, we concluded that differential responsiveness to TGFβ was to a certain extent correlated with expression of the transformed phenotype and that it could not be explained solely on the basis of altered production of endogenous TGFβ or to differences in TGFβ receptor binding characteristics. Rather, modulation of TGFβ inhibition of these human breast epithelial cells apparently occurs at a level distal to the TGFβ receptor.

In many other systems, such as in human bronchial epithelial cells, growth inhibition by TGFβ is parallelled by an induction of differentiation (Masui *et al.*, 1986). The effect of TGFβ on the expression of epithelial membrane antigen, a derivative of the breast epithelial specific marker milk fat globule protein (Stampfer and Bartley, 1988), was therefore studied in the 184 cells and sublines. A 10 to 15-fold increase in milk fat globule protein was induced in the normal cells following TGFβ treatment, while only a 2 to 3-fold increase was observed in the 184A1N4-T and 184A1N4-TH cells (Walker-Jones *et al.*, 1988). Sodium butyrate, another differentiation-inducing agent, was also found to stimulate milk fat globule protein expression. Thus, the growth inhibitory action of TGFβ on the human mammary epithelial cell lines is generally parallelled by induction of cellular differentiation, at least as measured by the breast epithelium specific marker milk fat globule protein.

It has been postulated that in fibroblasts, the mitogenic action of TGFβ is coupled to, or mediated by, an induction of expression of c-sis, the platelet-derived growth factor B-chain (PDGF, Leof *et al.*, 1987). We have previously reported the production of both the A and B chains of PDGF by human breast cancer cells (Bronzert *et al.*, 1987). In examining three of the normal human mammary epithelial cell strains, we found that they produced PDGF receptor competing activity in amounts similar to the levels determined in the conditioned media from the human breast cancer cells (Bronzert *et al.*, 1988). RNase protection analysis of total cellular RNA revealed strong expression of the PDGF A chain. Expression of the PDGF B chain was reversibly induced by TGFβ treatment, while PDGF A chain mRNA was not affected. At this point, the mechanism(s) behind this differential induction of PDGF B chain concomitant with growth inhibition by TGFβ in the human mammary epithelial cells, in contrast to the concomitant growth stimulation in fibroblasts, remain(s) to be determined.

CONCLUSIONS

From our studies of normal and oncogene transformed human mammary epithelial cells, we can conclude that some of the original hypotheses about mechanisms for malignant transformation, based on studies using fibroblasts, may need to be revised with respect to mammary carcinogenesis. TGFα, initially thought to be directly associated with the transformed state, may be more of a proliferation marker. This is also supported by the finding of immunoreactive TGFα and TGFα mRNA in benign human breast lesions (Perroteau *et al.*, 1986; Travers *et al.*, 1988). Likewise, in chemically induced rat mammary adenocarcinomas, the primary lesions were found to express TGFα whereas the serially transplantable carcinomas expressed little or no TGFα (Liu *et al.*, 1987). This lends additional support to the contention that enhanced TGFα production per se can not solely account for the transformed phenotype in mammary malignacies. Similar conclusions are drawn in a recent study of mouse epidermal or papilloma cells expressing a human TGFα cDNA in skin grafts on mice. TGFα produced by either tumor cells or adjoining normal cells could stimulate tumor growth, but TGFα could not directly influence tumor progression (Finzi *et al.*, 1988).

The notion that neoplastic proliferation involves escape from normal growth regulation such as inhibition by TGFβ is, at least partially, supported by our findings that the most fully transformed, tumorigenic member of the 184-derived cells are also least responsive to TGFβ. However, it is clear that the presence of activated oncogenes in these cells causes more profound perturbances in cell growth control than can be entirely explained by altered sensitivity to or production of TGFβ or TGFα.

Similarly, since oncogenic transformation only caused apparently minor changes in EGF receptor expression and no differences in TGFβ receptor expression in these human mammary epithelial cells, we can conclude that modulation of cellular responsiveness to these growth factors can occur at a level different from direct ligand-receptor interactions. Many questions regarding receptor functionality and second-messenger mechanisms have not yet been addressed in our studies of these cells.

In summary, we can clearly observe the effects of malignant transformation on cellular responsiveness to two of the major growth regulatory peptides, TGFα and TGFβ, in a model system of normal, immortalized, and oncogene transformed human mammary epithelial cells. It is apparent, however, that alterations of cellular responsiveness are effects of perturbations of growth control at a level beyond alterations in TGFα or TGFβ production or cell surface receptors. It will be the goal of our future endeavors to attempt to bring more light on the cause(s) or mechanisms of human mammary carcinogenesis.

REFERENCES

Anzano M.A., Roberts A.B., De Larco J.E., Wakefield L.M., Assoian R.K., Roche N.S., Smith J.M., Lazarus J.E., and Sporn M.B. 1985. Increased secretion of type ß transforming growth factor accompanies viral transformation of cells. Mol. Cell. Biol. 5:242.

Anzano M.A., Roberts A.B., Smith J.M., Sporn M.B., and DeLarco J.E. 1983. Sarcoma growth factor from conditioned medium of virally transformed cells is composed of both type α and type ß transforming growth factors. Proc. Natl. Acad. Sci. (USA) 80:6264.

Artega C.L., Hanauske A.R., Clark G.M., Osborne C.K., Hazarika P., Pardue R.L., Tio F., and Von Hoff D.D. 1988. Immunoreactive α transforming growth factor (IrαTGF) activity in effusions from cancer patients: a marker of tumor burden and patient prognosis. Cancer Research 48:5023.

Artega C.L., Tandon A.K., Von Hoff D.D., and Osborne C.K. 1988. Transforming growth factor ß: potential autocrine growth inhibitor of estrogen receptor-negative human breast cancer cells. Cancer Research 48:3898.

Assoian R.K., Grotendorst G.R., Miller D.M., and Sporn M.B. 1984. Cellular transformation by coordinated action of three peptide growth factors from human platelets. Nature (London) 309:804.

Assoian R.K., Komoriya A., Meyers C.A., Smith D.M., and Sporn M.B. 1983. Transforming growth factor ß in human platelets: identification of a major storage site, purification and characterization J. Biol. Chem. 259: 9756.

Balk S.D., Riley T.M., Gunther H.S., and Morisi A. 1985. Heparin-treated, v-myc-transformed chicken heart mesenchymal cells assume a normal morphology but are hypersensitive to epidermal growth factor (EGF) and brain fibroblast growth factor (bFGF); cells transformed by the v-Ha-ras oncogene are refractory to EGF and bFGF but are hypersensitive to insulin-like growth factors Proc. Natl. Acad. Sci. (USA) 82:5781.

Bates S.E., Davidson N.E., Valverius E.M., Dickson R.B., Freter C.E., Tam J.P., Kudlow J.E., Lippman M.E., and Salomon D.S. 1988a. Expression of transforming growth factor α and its mRNA in human breast cancer: its regulation by estrogen and its possible functional significance. Mol. Endocrinol. 2:543.

Bates S.E., McManaway M.E., Lippman M.E., and Dickson R.B. 1986. Characterization of estrogen responsive transforming activity in human breast cancer cell lines. Cancer Research 46:1707.

Bates S.E., Valverius E.M., Stampfer M., Dickson R.B., and Lippman M.E., 1988b. Transforming growth factor α production by proliferating normal breast epithelial cells. Proc. Am. Ass. Ca. Res. 29:52 (abstract 208).

Bronzert D.A., Pantazis P., Antoniades H.N., Kasid A., Davidson N., Dickson R.B., and Lippman M.E. 1987. Synthesis and secretion of PDGF-like growth factor by human breast cancer cell lines. Proc. Natl. Acad. Sci. (USA) 84:5763.

Bronzert D.A., Valverius E., Bates S.E., Stampfer M, and Dickson R.B. 1989. Production of α and ß chains of platelet derived growth factor by human mammary epithelial cells. 70th Annual Meeting of the Endocrine Society, Abstracts. in press.

Cheifetz S., Bassols A., Stanley K., Ohta M., Greenberger J., and Massague J. 1988. Heterodimeric Transforming Growth Factor ß. J. Biol. Chem. 263:10783.

Cherington P.V. and Pardee A.B. 1982. On the basis for loss of the EGF growth requirement by transformed cells. In "Growth of Cells in Hormonally Defined Media." (G.H. Sato, A.B. Pardee, D.A. Sirbasku, eds.), Cold Spring Harbor Conferences on Cell Proliferation 9, Cold Spring Harbor Laboratories, New York.

Ciardiello F., Kim N., Hynes N., Jaggo R., Redmond S., Liscia D.S., Sanfilippo B., Marlo G., Callahan R., Kidwell W.R., and Salomon D.S. 1988. Induction of transforming growth factor α expression in mouse mammary epithelial cells after transformation with a point-mutated c-Ha-ras protooncogene. Mol. Endocrinol. in press.

Clair T., Miller W.R., and Cho-Chung Y.S. 1987. Prognostic significance of the expression of a ras protein with a molecular weight of 21,000 by human breast cancer. Cancer Research 47:5290.

Clark R., Stampfer M.R., Milley R., O'Rourke E., Walen K.H., Kriegler M., Kopplin J., and McCormick F. 1988. Transformation of human mammary epithelial cells by oncogenic retroviruses. Cancer Research 48:4689.

Coffey R.J., Derynck R., Wilcox J.N., Bringman T.S., Goustin A.S., Moses H.L., and Pittelkow M.R. 1987. Production and auto-induction of transforming growth factor-α in human keratinocytes. Nature (London) 328:817.

Davidson N.E., Gelmann E.P., Lippman M.E., and Dickson R.B. 1987. Epidermal growth factor receptor gene expression in estrogen receptor-positive and negative human breast cancer cell lines. Mol. Endocrinol. 1:216.

Derynck R. 1988. Transforming growth factor α. Cell 54:593.

Di Fiore P.P., Pierce J.H., Kraus M.H., Segatto O., King C.R., and Aaronson S.A. 1987b. erbB-2 is a potent oncogene when overexpressed in NIH/3T3 cells. Science 237:178.

Di Fiore P.P., Pierce J.H., Fleming T.P., Hazan R., Ullrich A., King C.R., Schlessinger J., and Aaronson S.A. 1987a. Overexpression of the human EGF receptor confers an EGF-dependent transformed phenotype to NIH 3T3 cells. Cell 51:1063.

Dickson R.B., Kasid A., Huff K.K., Bates S., Knabbe C., Bronzert D., Gelman E.P., and Lippman M.E. 1987. Activation of growth factor secretion in tumorigenic states of breast cancer induced by 17-ß-estradiol or v-ras^{H} oncogene. Proc. Natl. Acad. Sci. (USA) 84:837.

Dickson R.B. and Lippman M.E. 1988. Control of human breast cancer by estroge, growth factors, and oncogenes, In "Breast Cancer: Cellular and Molecular Biology," (M.E. Lippman, R.B. Dickson, eds.), Martinus Nijhoff Publishers, Boston.

Dickson R.B., McManaway M.E., and Lippman M.E. 1986. Estrogen-induced factors of breast cancer cells partially replace estrogen to promote tumor growth. Science 232:1540.

Eckert K., Lübbe L., Schon R., and Grosse R. 1985. Demonstration of transforming growth factor activity in mammary epithelial tissues. Biochem. Intl. 11:441.

Fernandez-Pol J.A., Klos D.J., Hamilton P.D., and Talkad V.D. 1987. Modulation of epidermal growth factor receptor gene expression by transforming growth factor-ß in a human breast carcinoma cell line. Cancer Research 47:4260.

Filmus J., Pollak M.N., Cailleau R., and Buick R.N. 1985. MDA-468, a human breast cancer cell line with a high number of epidermal growth factor (EGF) receptors, has an amplified EGF receptor gene and is growth inhibited by EGF. Biochem. Biophys. Res. Commun. 128:898.

Finzi E., Fleming T., Segatto O., Pennington C.Y., Bringman T.S., Derynck R., and Aaronson S.A. 1987. The human transforming growth factor type α coding sequence is not a direct-acting oncogene when overexpressed in NIH 3T3 cells. Proc. Natl. Acad. Sci. (USA) 84:3733.

Finzi E., Kilkenny A., Strickland J.E., Balaschak M., Bringman T., Derynck R., Aaronson S., and Yuspa S.H. 1988. TGFα stimulates growth of skin papillomas by autocrine and paracrine mechnisms but does not cause neoplastic progression. Mol. Carinogenesis 1:7.

Goustin A.S., Leof E.B., Shipley G.D., and Moses L. 1986. Growth factors and cancer. Cancer Research 46:1015.

Hammond S.L., Ham R.G., and Stampfer M.R. 1984. Serum-free growth of human mammary epithelial cells: rapid clonal growth in defined medium and extended serial passage with pituitary extract. Proc. Natl. Acad. Sci. (USA) 81:5435.

Heldin C.H. and Westermark B. 1984. Growth factors: mechanism of action and relations to oncogenes. Cell 37:9.

Horan-Hand P., Vilase V., Thor A., Ohuchi N., and Schlom J. 1987. Quantitation of Harvey ras p21 enhanced expression in human breast and colon carcinomas. J. Natl. Cancer Inst. 79:59.

Houck K.A., Strom S.C., and Michalopoulos G. 1987. Resistance to the growth inhibitory effect of transforming growth factor β is induced by transfection of an activated H-ras oncogene into rat liver epithelial cells. Proc. Am. Ass. Ca. Res. 28:64 (abstract 254).

Hudziak R.M., Schlessinger J., and Ullrich A. 1987. Increased expession of the putative growth factor receptor $p185^{HER2}$ causes transformation and tumorigenesis of NIH 3T3 cells. Proc. Natl. Acad. Sci. (USA) 84:7159.

Ignotz R.A. and Massague J. 1986. Transforming growth factor-ß stimulates the expression of fibronectin and collagen and their incorporation into the extracellular matrix. J. Biol. Chem. 261:4337.

Kamata T. and Feramisco J.R. 1984. Is the ras oncogene protein a component of the epidermal growth factor receptor system? In "Cancer Cells 1/The Transformed Phenotype." (A.J. Levine, G.F. Vande Woude, W.C. Topp, J.D. Watson, eds.), Cold Spring Harbor, New York.

Kaplan P.L. and Ozanne B. 1983. Cellular responsiveness to growth factors correlates with a cell's ability to express the transformed phenotype, Cell 33:931.

Kimchi A., Wang X.-F., Weinberg R.A., Cheifetz S., and Massague J. 1988. Absence of TGF-β receptors and growth inhibitory responses in retinoblastoma cells. Science 240:196.

Kozma S.C., Bogaard M.E., Buser K., Saurer S.M., Bos J.L., Groner B., and Hynes N.E. 1988. The human c-Kirsten ras gene is activated by a novel mutation in codon 13 in the breast carcinoma cell line MDA-MB 231. Nucleic Acids Research 15:5963.

Kraus M.H., Yuspa Y., and Aaronson S.A. 1984. A position 12-activated H-ras oncogene in all Hs578T mammary carcinosarcoma cells but not normal mammary cells of the same patient. Proc. Natl. Acad. Sci. (USA) 81:5384.

Kurachi H., Okamoto S., and Oka T. 1985. Evidence for the involvement of the submandibular gland epidermal growth factor in mouse mammary tumorigenesis. Proc. Natl. Acad. Sci. (USA) 81:5940.

Leof E.B., Proper J.A., Shipley G.D., Di Corleto P.E., and Moses H.L. 1986. Induction of c-sis mRNA and activity similar to platelet-dervied growth factor by transforming growth factor-β: a proposed model for indirect mitogenesis involving autocrine activity. Proc. Natl. Acad. Sci. (USA) 83:2453.

Leof E.B., Proper J.A., and Moses H.L. 1987. Modulation of transforming growth factor type ß action by activated ras and c-myc. Mol. Cell. Biol. 7:2649.

Like B. and Massague J. 1986. The antiproliferative effect of type ß transforming growth factor ocurs at a level distal from receptors for growth-activating factors. J. Biol. Chem. 261:13426.

Linsley P.S., Hargreaves W.R., Twardzik D.R., and Todaro G.J. 1985. Detection of larger polypeptides structurally and functionally related to type I transforming growth factor. Proc. Natl. Acad. Sci. (USA) 82:356.

Liu S.C., Sanfilippo B., Perroteau I., Derynck R., Salomon D.S., and Kidwell W.R. 1987. Expression of transforming growth factor α (TGFα) in differentiated rat mammary tumors: estrogen induction of TGFα production. Mol. Endocrinol. 1:683.

Luttrell D.K., Luttrell L.M., and Parsons S.J. 1988. Augmented mitogenic responsiveness to epidermal growth factor in murine fibroblasts that overexpress pp60$^{c\text{-}src}$. Mol. Cell. Biol. 8:497.

Macias A., Perez R., Hägerström T., and Skoog L. 1987. Identification of transforming growth factor α in human primary breast carcinomas. Anticancer Research 7:1271.

Massague J. 1983. Epidermal growth factor-like transforming growth factor. J. Biol. Chem. 258:13606.

Masui T., Wakefield L.M., Lechner J.F., La Veck M.A., Sporn M.B., and Harris C.C. 1986. Type ß transforming growth factor is the primary differentiation-inducing serum factor for normal human bronchial epithelial cells. Proc. Natl. Acad. Sci. (USA) 83:2438.

Medina D. 1988. The preneoplasti state in mouse mammary tumorigenesis. Carcinogenesis 9:1113.

Moses H.L., Tucker R.F., Leof E.B., Coffey R.J., Halper J., and Shipley G.D. 1985. Type-ß transforming growth factor is a growth stimulator and a growth inhibitor. In "Cancer Cells" Vol.3, (J. Feramisco, B. Ozanne, C. Stiles, eds.), Cold Spring Harbor Laboratory, New York.

Oka T., Tsutsumi O., Kurachi H., and Okamoto S. 1988. The role of epidermal growth factor in normal and neoplastic growth of mouse mammary epithelial cells. In "Breast Cancer: Cellular and Molecular Biology." (M.E. Lippman and R.B. Dickson, eds.), Martinus Nijhoff Publishers, Boston.

Osborne C.K., Hamilton B., Titus G., and Livingston R.B. 1980. Epidermal growth factor stimulation of human breast cancer cells in culture. Cancer Research 40:2361.

Perez R., Pascual M., Macias A., and Lage A. 1984. Epidermal growth factor receptors in human breast cancer. Breast Cancer Re. Treat. 4:189.

Perroteau I., Salomon D., DeBortoli M., Kidwell W., Hazarika P., Pardue R., Dedman J., and Tam J. 1986. Immunological detection and quantitation of α transforming growth factors in human breast carcinoma cells. Breast Cancer Res. Treat. 7:201.

Reddel R.R., Ke Y., Kaighn M.E., Malan-Shibley L., Lechner J.F., Rhim J.S., and Harris C.C. 1989. Human bronchial epithelial cells neoplastically transformed by v-Ki-ras: altered response to inducers of terminal squamous differentiation. Oncogene Research. in press.

Riedel H., Massoglia S., Schlessinger J., and Ullrich A. 1988. Ligand activation of overexpressed epidermal growth factor receptors transforms NIH 3T3 mouse fibroblasts. Proc. Natl. Acad. Sci. (USA) 85:1477.

Roberts A.B., Anzano M.A., Wakefield L.M., Roche N.S., Stern D.F., and Spron M.B. 1985. Type ß transforming growth factor: a bifunctional regulator of cellular growth. Proc. Natl. Acad. Sci. (USA) 82:119.

Roberts A.B., Lamb L.C., Newton D.L., Sporn M.B., DeLarco J.E., and Todaro G.J. 1980. Transforming growth factors: Isolation of polypeptides from virally and chemically transformed cells by acid/ethanol extraction. Proc. Natl. Acad. Sci. (USA) 77:3494.

Rosenthal A., Lindquist P.B., Bringman T.S., Goeddel D.V., and Derynck R. 1986. Expression in rat fibroblasts of a human transforming growth factor-α cDNA results in transformation. Cell 46:301.

Sainsbury J.R., Farndon J.R., Needham G.K., Malcolm A.J., and Harris A.L. 1987. Epidermal-growth-factor receptor status as predictor of early recurrence of and death from breast cancer. Lancet i:1398.

Sairenji M., Suzuki K., Murakami K., Motohashi H., Okamoto T., and Umeda M. 1987. Transforming growth factor activity in pleural and peritoneal effusions from cancer and non-cancer patients. Jpn. J. Cancer Res. (Gann). 78:814.

Salomon D.S., Bano M., and Kidwell W.R. 1986. Polypeptide growth factors and the growth of mammary epithelial cells. In "Breast Cancer: Origins, Detection and Treatment," (M.A. Rich, J.C. Hayes, J.T. Papadimitrious, eds.), Martinus Nijhoff Publishers, Boston.

Salomon D.S., Perroteau I., Kidwell W.R., Tam J., and Derynck R. 1987. Loss of growth responsiveness to epidermal growth factor and enhanced production of α-transforming growth factors in ras-transformed mouse mammary epithelial cells. J. Cell. Physiol. 130:397.

Salomon D.S., Zwiebel J.A., Bano M., Losonczy I., Fehnel P., and Kidwell W.R. 1984. Presence of transforming growth factors in human breast cancer cells. Cancer Research 44:4069.

Santon J.B., Cronin M.T., MacLeod C.L., Mendelsohn J., Masui H., and Gill G.N. 1986. Effects of epidermal growth factor receptor concentration on tumorigenicity of A431 cells in nude mice. Cancer Research 46:4701.

Seyedin S.M., Thompson A.Y., Bentz H., Rosen D.M., McPherson J.M., Conti A., Siegel N.R., Gallupi G.R., and Piez K.A. 1986. Cartilage-inducing factor A. J. Biol. Chem. 261:5693.

Shankar V., Ciardiello F., Kim N., Derynck R., Liscia D.S., Merlo G., Langton B.C., Sheer D., Callahan R., Bassin R.H., Lippman M.E., Hynes N., and Salomon D.S. 1989. Transformation of normal mouse mammary epithelial cells following transfection with a human transforming growth factor α cDNA. Mol. Carcinogenesis submitted.

Shipley G.D., Pittelkow M.R., Wille J.J.Jr., Scott R.E., and Moses H.L. 1986 Reversible inhibition of normal human prokeratinocyte proliferation by type ß transforming growth factor-growth inhibitor in serum-free medium. Cancer Research 46:2068.

Silberstein G.B. and Daniel C.W. 1987. Reversible inhibition of mammary gland growth by transforming growth factor-ß. Science 237:291.

Slamon D.J., Clark G.M., Wong S.G., Levin W.J., Ullrich A., and McGuire W.L. 1987. Human breast cancer: correlation of relapse and survival with amplification of the HER-2/neu oncogene. Science 235:117.

Spitzer E., Grosse R., Kunde D., and Schmidt H.E. 1987. Growth of mammary epithelial cells in breast-cancer biopsies correlates with EGF binding. Int. J. Cancer 39:279.

Sporn M.B. and Roberts A.B. 1986. Peptide growth factors and inflammation, tissue repair, and cancer. J. Clin. Invest. 78:329.

Sporn M.B. and Roberts A.B. 1985. Autocrine growth factors and cancer. Nature (London) 313:745.

Sporn M.B. and Todaro G.J. 1980. Autocrine secretion and malignant transformation of cells. N. Engl. J. Med. 303:878.

Stampfer M.R. 1985. Isolation and growth of human mammary epithelial cells. J. Tiss. Cult. Meth. 9:107.

Stampfer M.R. and Bartley J.C. 1988. Human mammary epithelial cells in culture: differentiation and transformation. In "Breast Cancer: Cellular and Molecular Biology," (M.E. Lippman, R.B. Dickson, eds.), Martinus Nijhoff Publishers, Boston.

Stampfer M.R. and Bartley J.C. 1985. Induction of transformation and continuous cell lines from normal human mammary epithelial cells after exposure to benzo-a-pyrene. Proc. Natl. Acad. Sci. (USA) 82:2394.

Stern D.F., Hare D.L., Cecchini M.A., and Weinberg R.A. 1987. Construction of a novel oncogene based on synthetic sequences encoding epidermal growth factor. Science 235:321.

Stern D.F., Roberts A.B., Roche N.S., Sporn M.B., and Weinberg R.A. 1986. Differential responsiveness of myc- and ras- transfected cells to growth factors: selective stimulation of myc-transfected cells by epidermal growth factor. Mol. Cell. Biol. 6:870.

Stromberg K., Hudgins R., and Orth D.N. 1987. Urinary TGFs in neoplasia: immunoreactive TGF-α in the urine of patients with disseminated breast carcinoma. Biochem. Biophys. Res. Comm. 144:1059.

Takehara K., LeRoy E.C., and Grotendorst G.R. 1987. TGF-β inhibition of endothelial cell proliferation: alteration of EGF binding and EGF-induced growth-regulatory (competence) gene expression. Cell 49:415.

Todaro G.J., DeLarco J.E., and Cohen S. 1976. Transformation by murine and feline sarcoma viruses specifically blocks binding of epidermal growth factor to cells. Nature (London) 264:26.

Todaro G.J., Marquardt H., Twardzik D.R., Reynolds F.H., and Stephenson J.R. 1983. Transforming growth factors produced by viral-transformed and human tumor cells. In "Genes and Proteins in Oncogenesis," (I.B. Weinstein, H.J. Vogel, eds.), Academic Press, New York.

Travers M.R., Barrett-Lee P.J., Berger U., Luqmani Y.A., Gazet J-C., Powles T.J., and Coombes R.C. 1988. Growth factor expression in normal, benign, and malignant breast tissue. Brit. Med. J. 296:1621.

Valverius E.M., Bates S.E., Stampfer M.R., Clar R., McCormick F., Salomon D.S., Lippman M.E., and Dickson R.B. 1989. Transforming growth factor α production and EGF receptor expression in normal and oncogene transformed human mammary epithelial cells. Mol. Endocrinol. in press.

Valverius E.M., Walker-Jones D., Bates S.E., Stampfer M.R., Clark R., McCormick F., Lippman M.E., and Dickson R.B. 1988. Inhibition of human breast epithelial cells with transforming growth factor β and desensitization by oncogene mediated transformation. Proc. Am. Ass. Ca. Res. 29:238 (abstract 948).

Velu T.J., Beguinot L., Vass W.C., Willingham M.C., Merlino G.T., Pastan I., and Lowy D.R. 1987. Epidermal growth factor-dependent transformation by a human EGF receptor proto-oncogene. Science 238:1408.

Vonderhaar B.K. 1988. Regulation of development of the normal mammary gland by hormones and growth factors. In "Breast Cancer: Cellular and Molecular Biology," (M.E. Lippman and R.B. Dickson, eds.), Martinus Nijhoff Publishers, Boston.

Walker-Jones D., Valverius E.M., Stampfer M.S., Clark R., McCormick F., Lippman M.E., and Dickson R.B. 1988. Transforming growth factor β stimulates expression of milk fat globule protein in normal and oncogene-transformed human mammary epithelial cells. Proc. Am. Ass. Ca. Res. 29:249 (abstract 990).

Watanabe S., Lazar E., and Sporn M.B. 1987. Transformation of normal rat kidney (NRK) cells by an infectious retrovirus carrying a synthetic rat type α transforming growth factor gene. Proc. Natl. Acad. Sci. (USA) 84:1258.

INTRA-MAMMARY STEROID TRANSFORMATION: IMPLICATIONS FOR TUMORIGENESIS AND NATURAL PROGRESSION

W.R. Miller

University Department of Surgery
Royal Infirmary
Edinburgh, EH3 9YW, Scotland

Steroid hormones are heavily implicated in the development and continued growth of breast cancer. Castration, especially early in life has a major impact in reducing risk of breast cancer (Feinleib, 1968), a protection which may be abolished if replacement steroids are administered. Similarly, it has long been recognized that therapy based on steroid hormone deprivation may be associated with regression of established breast cancer. The presence of estrogen receptors in hormone dependent cancers (McGuire *et al*., 1975) and the effects of estrogen and anti-estrogen in cultures of breast cancer cells (Lippmann *et al*., 1983) point to the central role of estradiol amongst the steroid hormones. More recent work (Lippman *et al*., 1986) has suggested that the growth promoting properties of estrogen may at least in part be due to their ability to stimulate tumor cell secretion of polypeptide factors which possess the potential to trigger cell division in an autocrine or paracrine fashion (Figure 1A). It has also been suggested that hormone independent growth simply reflects an estrogen by-pass resulting from cellular adaptation towards the constitutive production of the same or similar growth factors (Dickson and Lippman, 1986).

However, with the observation that the breast and its tumors have the ability to synthesize active estrogen from precursor steroid (Miller, 1984), it is possible to envisage a second form of autocrine/paracrine control, especially after the menopause when the gonadal contribution of estrogen to the breast is much reduced. In this scheme (Figure 1B), estrogen biosynthesis within the breast maintains estrogen-dependent processes within mammary tumors, and the role of polypeptide factors would be to control this local steroidogenesis.

The evidence to support this concept will be assessed by reviewing the potential of the breast for estrogen biosynthesis and metabolism, the evidence that estrogen biosynthesis may be an autocrine factor in breast cancer, the complexity of the breast and the paracrine inter-relationships between different compartments and, finally, the data which suggest that steroid metabolism in breast adipose tissue may be regulated by breast cancer-derived paracrine factors, which may have a similar action to defined growth factors.

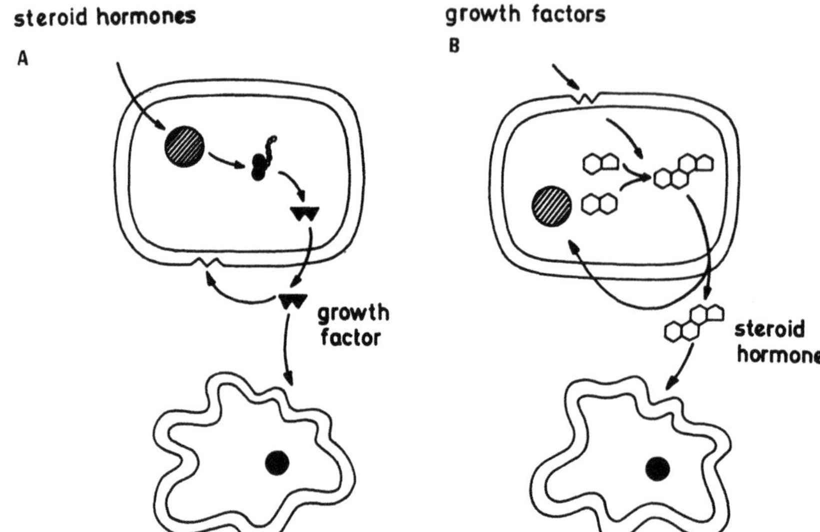

Figure 1. Potential autocrine and paracrine effects mediated by polypeptide growth factors and steroid hormones. In scheme A, steroid hormones induce the secretion of growth factors which exert autocrine and paracrine action. In scheme B growth factors regulate the synthesis of steroids which can then act to maintain events within the cell of synthesis (autocrine) or be secreted to influence neighboring cells (paracrine). It is also possible to envisage a hybrid paracrine scheme in which growth factors secreted locally induce steroid biosynthesis in adjacent cells.

STEROID METABOLISM IN THE BREAST AND ITS TUMORS

Whereas steroid hormones, which may influence the growth and development of both normal and diseased breasts, are classically synthesized by the gonads, it is now recognized that other organs including the breast and its tumors are capable of steroidogenesis (Miller, 1984). Many of the steroid transformations in the breast are associated with marked changes in the biological activity between precursor and product. Thus, the potential exists for paracrine and autocrine activity modulated by such steroid metabolism. In view of the central role of estrogens in the breast, two important steroid metabolizing enzymes as shown in Figure 2 are aromatase which catalyses the conversion of C_{19} androgens into C_{18} estrogens with the concomitant aromatization of the steroid A ring and 17β-hydroxysteroid dehydrogenase(s) which transforms the relatively inactive estrone into the more potent estrogen, estradiol. The potential importance of these enzymes is reinforced by the relatively high concentrations of androgens and estrone within the breast and circulation even in postmenopausal women.

Although aromatase has never been convincingly shown in benign breast tumors and normal parenchymal cells, the activity appears to be present in breast adipose tissue and 60-70% of breast cancers (Miller, 1986). In terms of paracrine activity it may be important that different clones of malignant cells from the same tumor may display different capacities for aromatization (Perel *et al.*, 1983). Levels of aromatase in breast cancer are small in comparison with those in placenta but are similar or higher than those in other peripheral organs (Abul-Hajj *et al.*, 1979; Perel *et al.*, 1980).

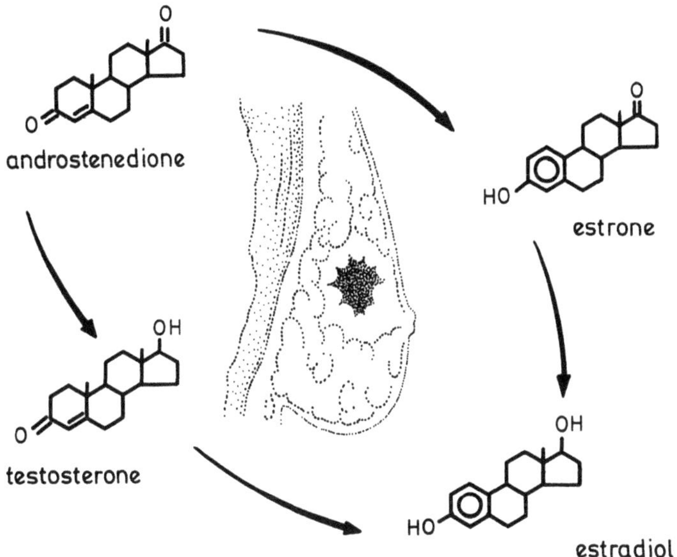

Figure 2. Steroid metabolism in the breast showing the conversion of androstenedione to estrone and testosterone to estradiol by aromatization and of androstenedione to testosterone and estrone to estradiol by 17β-hydroxysteroid dehydrogenase activity.

Seventeen β-hydroxysteroid dehydrogenase is found in all types of breast tissue although controversy exists as to whether activity is higher or lower in cancer as compared with non-malignant material (Bonney *et al.*, 1983). There may also be several forms of the enzyme, all catalysing the interconversion between estrone and estradiol but differing in their co-factor requirements, kinetics of reactions, subcellular and tissue distribution and their predominate direction of reaction (Pollow *et al.*, 1977). The latter is an issue of some importance since estradiol has a greater biological potency as is reflected by its high affinity for the estrogen receptor. Transformation of estrone to estradiol will therefore increase estrogenicity whereas the reverse reaction is a pathway of deactivation. However, the position of equilibrium is not clear; *in vitro* studies strongly suggest that the balance is in favor of the conversion to estradiol to estrone (Bonney *et al.*, 1983), but results from *in vivo* perfusion (McNeil *et al.*, 1986a) and comparison of endogenous steroids with those in the circulation (Van Landegham *et al.*, 1985) indicate the predominance of estrone to estradiol.

While controlling factors for aromatase in the ovaries (Channing and Segal, 1981) and 17 β-hydroxysteroid dehydrogenase in the endometrium (Buirchell and Hahnel, 1979) have been defined, comparatively little is known about regulation of the enzymes in the breast. This complicates the assessment of these enzymes as paracrine/autocrine agents.

STEROID METABOLISM AS AN AUTOCRINE FACTOR IN BREAST CANCER

The basic question to be addressed is whether certain breast cancers can synthesize sufficient estrogen by means of their own metabolic activity to maintain estrogen dependent events within the tumor.

Unfortunately, the definitive answer is not available, and it is only possible at this time to review the evidence against and in favor of the proposition.

Again the notion that local steroid metabolism is of significance is the lack of any significant correlation between tumor enzyme activity (as measured *in vitro*) and endogenous levels of estrogen within the breast (Bonney *et al.*, 1983); Vermeulen *et al.*, 1986; Hawkins *et al.*, 1987). for example, work in our laboratories has failed to detect significant differences in concentrations of estradiol in tumors with and without aromatase activity and with differing 17 β-hydroxysteroid dehydrogenase activity (Miller and O'Neill, in press).

Preliminary data from studies of patients with breast cancer who have been perfused with radioactively-labelled steroids have produced more promising results. These suggest that a variable but substantial amount of estrogen within the breast comes from local metabolism (James, in press).

Table 1. Relationship between tumor aromatase and estrogen receptors

	Estrogen Synthesis	Without Synthesis
Receptor-positive	93	49
Receptor-negative	27	31

$X^2 = 6.16$, $p<0.025$

Our own approach has been to compare tumor steroidogenesis, in particular aromatization, with the clinical response of that tumor to steroid deprivation therapy. To understand the rationale behind this, it is helpful to look at the relationship between aromatase and estrogen receptors. Our own results are shown in Table 1. This shows that we, unlike other workers who have studied smaller numbers of tumors (Tilson-Mallet *et al.*, 1983; Li and Adams, 1981), have found a positive tendency for estrogen biosynthesis to be associated with estrogen receptor-positive tumors. However it is more important to emphasize two major points of agreement between the various studies and discuss their implications. These are that estrogen biosynthesis may be detected in estrogen receptor-positive and -negative tumors and that estrogen receptor-positive tumors may or may not have the potential for estrogen biosynthesis.

The significance of aromatase activity in estrogen receptor-negative tumors is debatable. Although it has been suggested that certain estrogen receptor-negative tumors only appear to be endocrine-independent because they can satisfy their hormonal requirement by means of their own biosynthetic capacity (Abul-Hajj, 1979), there are data which would argue against this. In particular, anti-estrogen therapy, which would be effective in hormone sensitive tumors irrespective of the source of estrogen, is rarely successful in estrogen receptor-negative tumors (Hawkins *et al.*, 1980). However, it seems more likely that, if estrogen biosynthesis is to be of significance, it will be in estrogen receptor positive tumors in which a mechanism exists to process the synthesized steroid. Indeed, if *in situ* synthesis is a major route by which certain breast cancers obtain their estrogen, then it might be expected that these tumors will respond to therapies which inhibit tumor aromatase.

Furthermore, since only a proportion of estrogen receptor tumors respond to anti-aromatase therapy (Lawrence *et al.*, 1980), it can be asked whether tumors possessing both estrogen receptors and aromatase activity are more likely to respond to this form of treatment than tumors with estrogen receptors but no aromatase activity.

Table 2. Relationship between tumor aromatase and response to aminoglutethimide therapy

	Aromatase Positive	No Aromatase
Clinical response	11	0
Disease progression	7	5

Fishers exact test $p = 0.047$

To address this question we have measured aromatase in tumors from postmenopausal women with advanced breast cancer. All patients had estrogen receptor-positive tumors and were treated with the aromatase inhibitor, aminoglutethimide. The relationship between tumor aromatase status and response to treatment is shown in Table 2. All five tumors without aromatase activity failed to respond whereas 11 of 18 with estrogen biosynthesis responded to therapy. The presence of aromatase in a tumor is therefore not an absolute marker for response to anti-aromatase therapy but it would seem that beneficial effects are more likely to be achieved in this group.

Since similar relationships have not been found between tumor aromatase and response to ovariectomy in premenopausal women or to tamoxifen in postmenopausal women, it would seem that the presence of aromatase is at least partially specific to anti-aromatase treatment and not merely a marker of endocrine sensitivity. Whilst the data would be compatible with remission following the blockage of tumor autocrine production of estrogen by aminoglutethimide, there is an alternative possibility of aromatase activity in breast cancers simply reflecting levels of estrogen biosynthesis in other tissues. Because aminoglutethimide's action is not restricted to tumor aromatase and the drug equally inhibits estrogen biosynthesis in, for example, adipose tissue, the supply of estrogen to the tumor via the circulation will also be correspondingly reduced.

RELATIONSHIPS BETWEEN DIFFERENT BREAST COMPARTMENTS AND BREAST CANCER

Because breast cancers mainly develop in lobular and ductal units, it is important to study relationships between normal and malignant epithelia. However, the breast is a complex organ in which, except during pregnancy, parenchyma constitutes only a relatively small component. As is shown in Figure 3, other elements such as stroma and adipose tissue form the bulk of the breast although their proportions may vary markedly at different stages of development and evolution (Preschtel, 1977). It can be seen, therefore, that the potential for interplay and paracrine regulation exists and consideration ought to be given to the roles of stroma and adipose tissue in the natural history and progression of breast cancer.

Although stroma within the breast has been traditionally thought of as structural scaffold. for more important elements such as ducts and

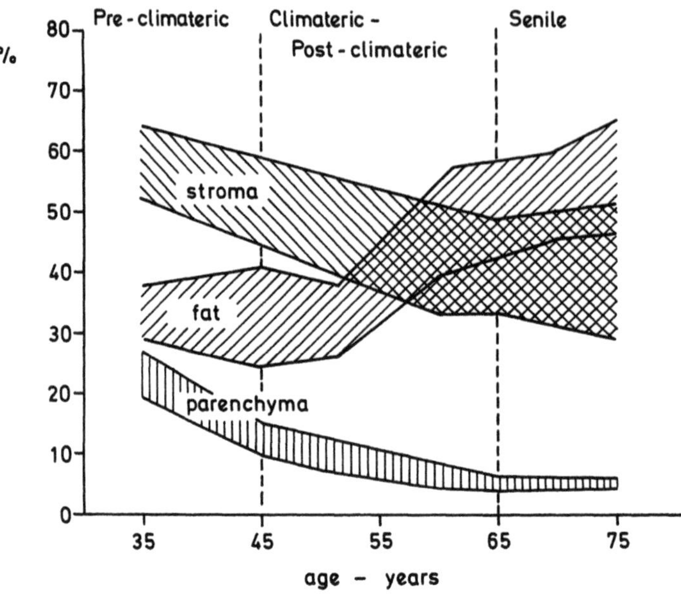

Figure 3. The proportion of the major elements in the female breast and their variation with age.

lobular-alveolar units, there is evidence that stroma plays a more active role (Tanzer and Spring-Mills, 1984) and that paracrine effects between the stromal and epithelial compartments are important in the development of both normal and malignant breast. In rodents, the differentiation and growth of the mammary epithelial buds may occur in response to steroids. These do not affect the epithelium but act through the mediation of the surrounding fibroblasts (Durnberger *et al.*, 1978; Haslem, 1986). A similar interaction between adjacent epithelial and mesenchymal tissues may also play a part in the control and development of the breast in the adult. As a result, it has been postulated that the continued presence of a fibroblast product may increase susceptibility to the development of epithelial tumors (Schor *et al.*, 1987). Certainly fibroblasts from patients with breast cancer may behave abnormally (Durning *et al.*, 1984). Data have also been presented that fibroblasts stimulate the growth of human breast cancer cells growing in immunosuppressed mice (Horgan *et al.*, 1987). Conversely, breast cancer cells may produce a growth factor like PDGF which is mitogenic for fibroblasts (Rozengurt *et al.*, 1985) .

The involvement of mammary adipose tissues has been relatively overlooked, although compared with other glands in the body, the adult human breast is unusual in being invested in an abundance of adipose tissue. This is most marked in older women, since the ratio of breast adipose tissue to glandular tissue increases with age (Preschtel, 1977). The role of mammary adipose tissue is not well understood but there is evidence of a link with the development of mammary cancer in both animals and humans. Thus, only animal species with well developed fat pads develop mammary cancer and, in rodents, animals with little mammary adipose tissue or experimentally-cleared fat pads appear resistant to mammary neoplasms. In mice, the mammary fat pad appears to be essential for the development of hyperplastic lesions and cancers (Deome *et al.*, 1959). Epidemiological evidence would support a similar role in humans. Obesity, which increases the amount of adipose tissue in the breast (Strombeck, 1964), is associated with an increased risk of breast cancer in postmenopausal women (De Waard, 1983). Similarly, the relative proportion of adipose tissue increases with age in parallel with the incidence of breast cancer. A comparison of Japanese immigrants to Hawaii with native Japanese women also showed a similar positive correlation between the proportion of mammary adipose tissue and the incidence of hyperplastic breast lesions and breast cancer (Sasano *et al.*, 1978).

There are several explanations to account for the involvement of mammary adipose tissue in breast cancer development. The fat within the adipose tissue may act as a sponge for organic carcinogens. Alternatively, breast adipose tissue is not metabolically inert and it has the potential to synthesize active steroid hormones which might promote the development of transformed cells (Deslypere *et al.*, 1985). This steroidogenesis and its paracrine regulation is thus worthy of further attention.

PARACRINE CONTROL OF STEROID METABOLISM IN BREAST ADIPOSE TISSUE

Aromatase activity may vary widely between different specimens of breast fat values. However, levels of aromatase are significantly higher in breast adipose tissue derived from breast cancer patients compared with that in fat from women with benign breast conditions (O'Neill and Miller, 1987). This would suggest that either the fat in which tumors develop is inherently different or tumors secrete factors which induce or stimulate aromatase activity.

These studies have been taken further by comparing aromatase activity in fat taken from the periphery of each quadrant of 12 consecutive mastectomy specimens from patients being treated for breast cancer. Correlations have then been made with anatomical derivation and tumor location. Considerable differences (up to five-fold) were detected between different quadrants of the same breast (O'Neill *et al.*, 1988). This variation in activity displayed some anatomical basis. For example, the upper outer quadrant had the highest activity in 7 of the 12 specimens and was never the site of lowest activity; in contrast, the lower inner quadrant never had the highest activity and was the source of the lowest activity in five cases. This correlation between aromatase and anatomical derivation was statistically significant ($p>0.05$).

However, the primary aim of this study was to relate aromatase activity to the site of tumor. These results are shown in Figure 4. In all twelve mastectomy specimens, the quadrant displaying the highest level of aromatase activity contained palpable tumor and, conversely, the quadrant with the lowest activity never contained tumor. Furthermore, in those breasts in which tumor was present in more than one quadrant, tumor-bearing quadrants always had higher aromatase than non-tumor-bearing quadrants. The relationship between tumor presence and level of aromatase was therefore absolute and statistically significant. Multiple regression analysis showed that, after correction for tumor presence, the correlation between tumor aromatase and anatomical derivation no longer reached statistical significance.

There are several potential explanations for the foregoing observations. First, as breast cancers invariably display higher aromatase activity than adipose tissue, samples taken in tumor-bearing quadrants might be more likely to contain microscopic deposits of tumor. This possibility cannot be totally excluded, but samples of fat adjacent to those taken for aromatase assay showed no evidence of gross

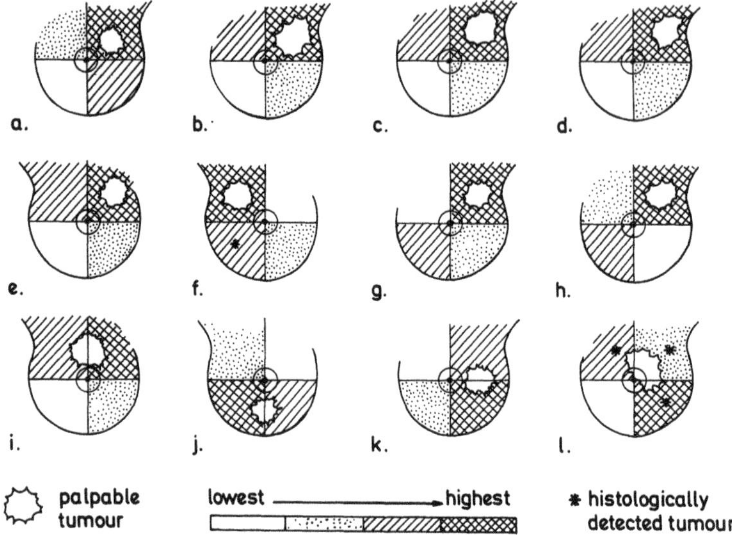

Figure 4. Aromatase activity in adipose tissue from breast quadrants ranked according to the shade scale and correlated with tumor location as indicated.

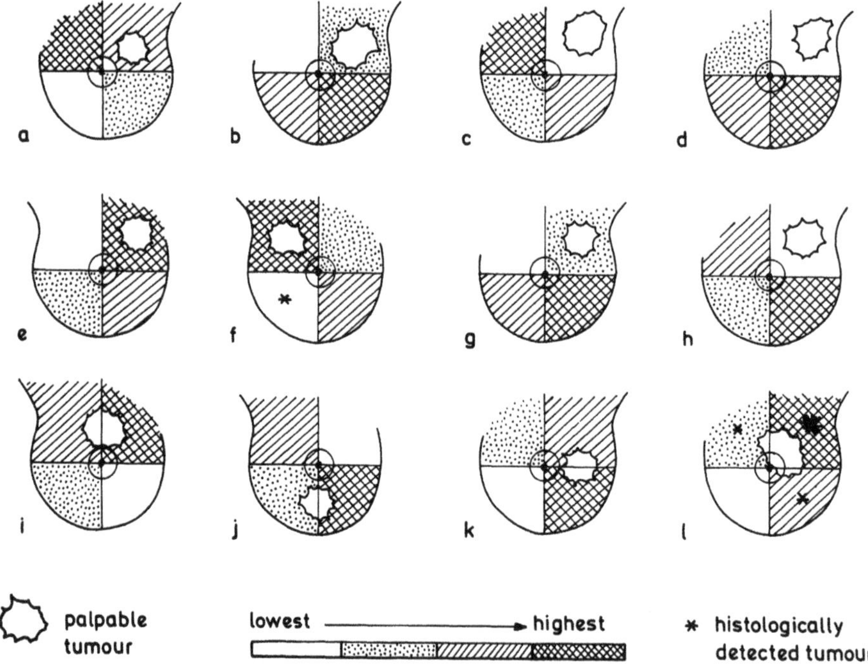

Figure 5. 17β-hydroxysteroid dehydrogenase activity in adipose tissue from the same breast quadrants as in Figure 4. Activities were ranked according to the shade scale and correlated with tumor location as indicated.

involvement. A second explanation for the results is that there is an inherent, but non-specific characteristic about fat found in proximity to cancer, since the cancers grow in the more cellular parts of the breast. This seems unlikely since the cytosol proteins of the fat specimens did not mirror the variation in aromatase activity. In addition, the activity of 17 β hydroxysteroid dehydrogenase in the same tissue samples did not correlate with either anatomical site or location of tumor (Figure 5), suggesting that enhanced aromatase is not a reflection of a general increase in androgen metabolism.

It is also possible that regionally increased aromatase activity preceded the appearance of the tumor. Indeed, it could be postulated that enhanced aromatase would lead to a locally high concentration of estrogen which in turn would encourage malignant growth at that particular site. This conjecture can neither be confirmed or refuted at this stage.

However, in the context of the present discourse, the possibility with the most relevance is that breast tumors secrete factors into their local environment which either induce or stimulate aromatase activity. In support of this it has been shown that breast cancer cells may secrete growth factors such as epidermal growth factor (EGF), transforming growth factors α and β (TGF-α and TGF-β; Lippman *et al.*, 1986) and that these factors have the potential to influence aromatase activity in adipose tissue (Simpson and Mendelson, 1987).

We also have preliminary data which would be consistent with paracrine secretion of tumor factors influencing steroidogenesis in adipose tissue cells. These results are summarized in Figure 6 and relate to a fibroblast (pre-adipocyte) cell line derived from breast fat. This cell line displayed aromatase activity which could be markedly induced by culturing in the presence of dexamethasone for 48 hours. The inclusion of EGF in the medium produced different effects dependent upon whether dexamethasone was present or not. In the absence of dexamethasone, EGF was stimulatory; in its presence, aromatase was inhibited by EGF. Of particular interest was the effect of including a homogenate of breast cancer for the 48 hour culture period in the same fibroblast cell line. Effects again were dependent upon whether dexamethasone was present. In its absence, the tumor homogenate caused an increase in fibroblast aromatase activity whereas in the presence of dexamethasone the homogenate was inhibitory. The effects of the addition of material derived from this breast cancer were similar to those of the growth factor. The inference is that the active principle within the tumor has properties compatible with the action of a growth factor.

Whilst there is no evidence that 17β-hydroxysteroid dehydrogenase activity is influenced by the presence of a breast tumor, there are data that suggest that levels of the enzyme in breast fat increase as breast cancers become more advanced. Thus, there is a positive association between tumor size and 17β-hydroxysteroid dehydrogenase activity (Beranek *et al.*, 1985). Our own results in Figure 7 show that adipose tissue associated with large tumors ($\geq$3cm) has significantly increased levels of activity compared with those in fat surrounding smaller tumors (>3cm). Similarly as is shown in Figure 8, levels of 17β-hydroxysteroid dehydrogenase activity were found to be significantly higher in breast fat from women whose cancers had spread to axillary lymph nodes as compared with activity in breast fat from those who were pathologically node-negative. We have also studied 17β-hydroxysteroid dehydrogenase activity in breast fat after systemic therapy given in an attempt to reduce the size of the primary tumor. These results (Figure 9) show that

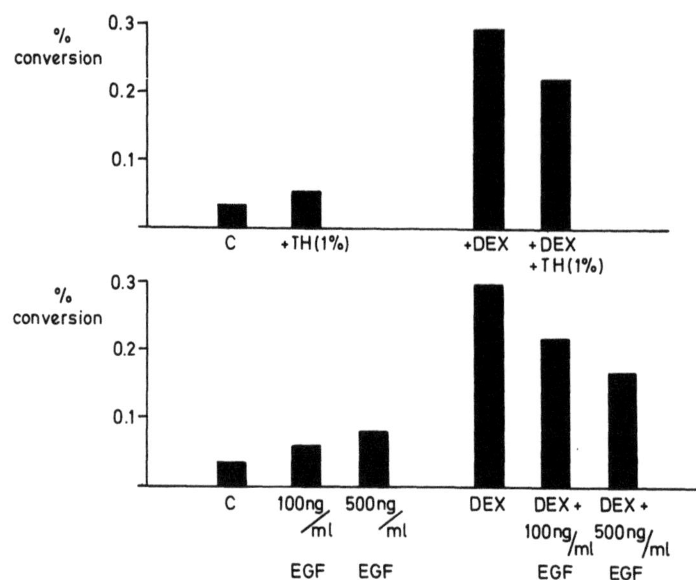

Figure 7. 17β-hydroxysteroid dehydrogenase (as measured by the conversion of androstenedione to testosterone) in breast adipose tissue from patients with small (<3cm) and large (≥3cm) breast cancers. The horizontal bars denote the median activity for the group. The p value is for the statistical difference between the groups as derived from the Wilcoxon rank test.

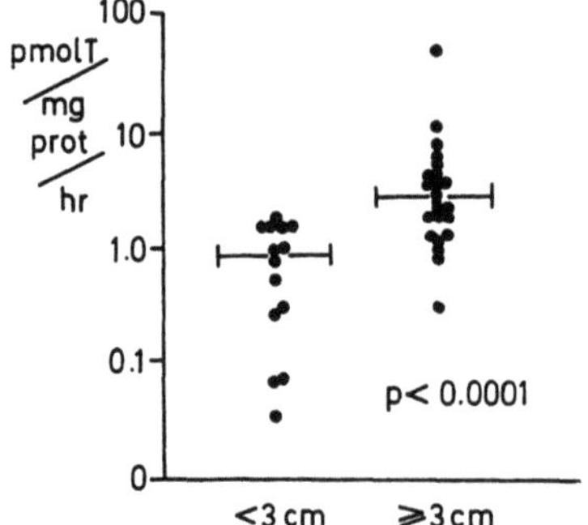

Figure 6. Aromatase activity (expressed as percentage conversion of androstenedione to estrogen) in human breast fat fibroblasts cultured in the absence of hormone additions (C), with a homogenate (1%) of breast cancer (TH), with dexamethasone (DEX), with both dexamethasone and the breast cancer homogenate (DEX + TH), with varying concentrations of epidermal growth factor (EGF) and with both dexamethasone and epidermal growth factor (DEX + EGF).

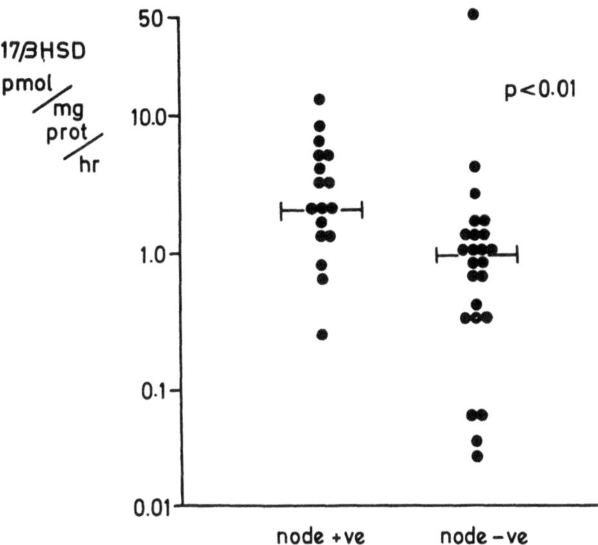

Figure 8. 17β-hydroxysteroid dehydrogenase in breast adipose tissue from breast cancer patients with and without nodal metastasis. Horizontal bars denote the median activity for the group. The p value is for the statistical difference between the groups as derived form the Wilcoxon rank test.

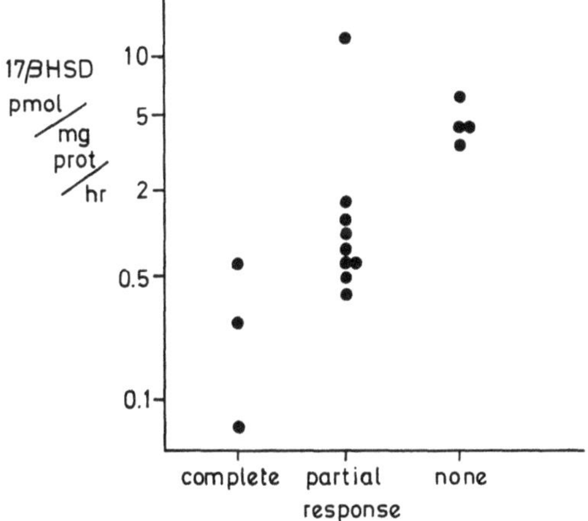

Figure 9. 17-β hydroxysteroid dehydrogenase activity in breast adipose tissue from patients with breast cancer following systemic therapy. The results are subdivided according to the response of the primary tumor to treatment ("complete response" represents clinical and pathological disappearance of tumor, "partial response" represents a decrease in clinical tumor size but the continued pathological presence of tumor, "none" represents an increase in clinical tumor size during treatment).

breast fat associated with tumors resistant to systemic therapy display a higher level of activity than that measured in breast fat surrounding tumors which responded to systemic therapy. In contrast, aromatase activity in breast fat was not associated with either tumor size, lymph node involvement or tumor response to systemic therapy.

These data are consistent in that 17β-hydroxysteroid dehydrogenase activity is elevated in breast fat associated with tumors of a more advanced stage (in terms of size and lymph node involvement) and poor prognosis (resistance to systemic therapy). Whether this relationship is casual or causal is unknown, but again the observations are compatible with the concept that more aggressive tumors secrete increased levels of different factors into their surrounding compartments compared with cancers of an earlier stage or of a less aggressive nature. It is therefore pertinent that homogenates of breast cancer and growth factors known to be secreted by tumors are able to affect 17β- hydroxysteroid dehydrogenase activity when added to cultures of breast fat (McNeill *et al.*, 1986b).

However since aromatase but not 17β-hydroxysteroid dehydrogenase is associated with the presence of tumor whereas dehydrogenase but not aromatase is associated with advancement of disease, it must be assumed that different factors are involved in the early and late stages of tumor progression.

CONCLUSIONS

The scheme shown in Figure 1A whereby steroid hormones stimulate cellular growth of tumors by inducing the increased secretion of mitogenic growth factors which act in an autocrine or paracrine manner is now established as a practical possibility. The complimentary scheme shown in Figure 1B whereby a polypeptide factor induces or stimulates steroid biosynthesis and metabolism by breast cells and the resultant estrogen promotes tumor development has received less attention and is consequently founded on less firm data. Nevertheless, there is good evidence that malignant and normal elements of the breast possess the potential to synthesize biologically active estrogen from less active precursors. The factors which control local steroidogenesis within the breast remain largely undefined although, at least in culture, certain polypeptides may modify activity in cells derived from breast adipose tissue.

Furthermore, there are certain positive relationships between steroid metabolism in the breast and the presence, stage and hormone dependence of breast cancer which would be compatible with autocrine and paracrine communication. These include: the correlation between tumor aromatase and clinical response to anti-aromatase therapy (This can be interpreted as evidence in favor of an autocrine loop whereby tumors satisfy their requirements for estrogen by means of their own biosynthetic capacity.); the relationship between the presence of breast cancer and enhanced aromatase activity in the surrounding breast fat. (This suggests that tumor and adipose tissue may communicate using a paracrine mechanism.); and finally, the correlation between enhanced 17β-hydroxysteroid dehydrogenase activity in breast fat and various aspects of tumor advancement and autonomy. (This reinforces the interrelationship between different components of the breast and suggest that the signals may vary between either tumors with different degrees of autonomy or those with differing stage.). The preliminary observations that polypeptide factors and tumor extracts may produce similar changes in steroid metabolism by breast tissue would also support the concept that intra-mammary steroidogenesis is a potentially important signalling system.

REFERENCES

Abul-Hajj Y.J. 1975. Metabolism of dehydroepiandrosterone by hormone dependent and independent human breast cancer. Steroids 26:488.

Abul-Hajj Y.J., Iverson R. and Kiang D.T. 1979. Aromatization of androgens by human breast cancer. Steroids 33:205.

Beranek P.A., Fokerd E.J., Newton C.J., Reed M.J., Chilchick M.W. and James V.H.T. 1985. The relationship between 17β-hydroxysteroid dehydrogenase and breast tumour site and size. Int. J. Cancer. 36:685.

Bonney R.C., Reed M.J., Davidson K., Beranek P.A. and James V.H.T. 1983. The relationship between 17β-hydroxysteroid dehydrogenase activity and oestrogen concentration in human breast tumours and in normal breast. Clin. Endocrinol. 19:727.

Buirchell B.J. and Hahnel R. 1985. Metabolism of estradiol-17β in human endometrium during the menstrual cycle. J. Steroid Biochem. 6:1489.

Channing C.P. and Segal S. 1981. Intraovarian control mechanisms. Plenum Press: New York.

Dickson R.B. and Lippman M.E. 1986. Hormonal control of human breast cancer cell lines. Cancer Surv. 5:617.

De Ome K.B., Franklin L.J. and Bern H.A. 1959. Development of mammary tumours from hyperplastic nodules transplanted into gland free mammary fat pads of female C3H mice. Cancer Research 19:515.

Deslypere J.P., Verdonck L. and Vermeulen A. 1985. Fat tissue: a steroid reservoir and site of steroid metabolism. J. Clin. Endocrinol. Metab. 61:564.

Durnberger H., Heuberger B., Schwartz P., Wasner G. and Kratochwil K. 1978. Mesenchyme-mediated effect of testosterone on embryonic mammary epithelium. Cancer Research 38:4066.

Durning P., Schor S.L. and Sellwood A.B. 1984. Fibroblasts from patients with breast cancer show abnormal migratory behaviours *in vitro*. The Lancet 2:890.

De Waard F. 1983. Epidemiology of breast cancer: a review. Eur. J. Cancer Clin. Oncol. 19:1671.

Feinleib M. 1968. Breast cancer and artificial menopause: A cohort study. J. Natl. Cancer Inst. 41:315.

Haslam S.Z. 1986. Mammary fibroblast influence on normal mouse mammary epithelial cell response to oestrogen *in vitro*. Cancer Research 46:310.

Hawkins R.A., Thijssen J.H.H. and Miller W.R. 1987. Die bedeutung der Aromatase-und Sulfatasestoffwechselwege fur die intrazellulare Ostrogensynthese beim Mammakarzmom postmenopausoler Frauen. Act. Onkol. 38:3.

Hawkins R.A., Roberts M.M. and Forrest A.P.M. 1980. Oestrogen receptors and breast cancer: current status. Br. J. Surg. 67:153.

Horgan K., Jones D.L. and Mansel R.E. 1987. Mitogenicity of human fibroblasts *in vivo* for human breast cancer cells. Br. J. Surg. 74:227.

James V.H.T. 1989. Oestrogen uptake and metabolism *in vivo*. Proc. Roy. Soc. (Edin). (in press).

Lawrence B.V., Lipton A., Harvey H.A., Santen R.J., Wells S.A., Cox C.E., White D.S. and Smart E.E. 1980. Influence of estrogen receptor status on response of metastatic breast cancer to aminoglutethimide therapy. Cancer 45:786.

Li K. and Adams J.B. 1981. Aromatization of testosterone and oestrogen receptors levels in human breast cancer. J. Steroid Biochem. 14:269.

Lippman M.E., Bolan G. and Huff K. 1983. The effects of estrogens and antiestrogens on hormone-responsive human breast cancer cell line. Cancer Research 43:1244.

McGuire W.L., Carbone P.P., Sears M.E. and Escher G.C. 1975. Estrogen receptors in human breast cancer: an overview. In: "Estrogen Receptors Human Breast Cancer", (W.L. McGuire, P.P. Carbone and E.P. Vollmer, eds.), Raven Press, New York, p.1.

McNeil J.M., Reed M.J., Beranek P.A., Bonney R.C., Chilchick M.W., Robinson D.J. and James V.H.T. 1986(a). A comparison of the *in vivo* uptake and metabolism of 3H-oestrone and 3H oestradiol by normal breast and breast tumour tissue in postmenopausal women. Int. J. Cancer 38:193.

McNeil J.M., Reed M.J., Beranek P.A., Newton C.J., Chilchik M.W. and James V.H.T. 1986(b). The effect of epidermal growth factor, transforming growth factor and breast tumour homogenates on the activity of oestradiol 17β- hydroxysteroid dehydrogenase in adipose tissue. Cancer Reports 31:213.

Miller W.R. and O'Neill J.S. 1989. The relevance of local oestrogen metabolism within the breast. Proc. Roy. Soc. (Edin). (in press).

O'Neill J.S. and Miller W.R. 1987. Aromatase activity in breast adipose tissue from women with benign and malignant breast disease. Br. J. Cancer 56:601.

O'Neill J.S., Elton R.A. and Miller W.R. 1988. Aromatase activity in adipose tissue from breast quadrants: a link with tumour site. Br. Med. J. 296:741.

Perel E., Wilkin D. and Killinger D.W. 1980. The conversion of androstenedione to estrone, estradiol and testosterone in breast tissue. J. steroid Biochem. 13:89.

Perel E., Blackstein M.E. and Killinger D.W. 1983. Control of aromatase activity in cultured human breast carcinoma cells. J. Steroid Biochem. 19: suppl 645 (Abs. 191).

Pollow K., Boquoi E., Banmann J., Schmidt-Gollwitzer M. and Pollow B. 1977. Comparison of the *in vitro* conversion of estradiol to estrone of normal and neoplastic human breast tissue. Mol. Cell. Endocrinol. 3:33.

Preschtel K. 1977. Benign disease of the female breast: histology, normal and abnormal. In: "Proceedings of the VIIIth World Congress on Gynaecology and Obstetrics", (L. Castelazo-Ayalah and C. MacGregor, eds.). p.135.

Rozengurt E., Sinnett-Smith J. and Taylor-Papadimitriou J. 1985. Production of PDGF-like growth factor by breast cancer cell lines. Int. J. Cancer 36:247.

Schor S.L., Schor A.M., Howell A. and Crowther D. 1987. Hypothesis: the persistent expression of fetal-like phenotypic characteristics by fibroblasts is associated with an increased susceptibility to neoplastic disease. Exp. Cell Biol. 55:11.

Simpson E.R. and Mendelson C.R. 1987. Effect of aging and obesity on aromatase activity of human adipose cells. Am. J. Clin. Nut. 45:290.

Strombeck J.O. 1944. Macromastia in women and its surgical treatment: a clinical study based on 1042 cases. Acta Chir. Scand. suppl 341:33.

Tanzer M.L. and Spring-Mills E. 1984. Breast cancer, epithelial cells and extracellular matrix. J. Natl. Cancer Inst. 73:999.

Tilson-Mallet N., Santner S.J., Feil P.D. and Santen R.J. 1983. Biological significance of aromatase activity in human breast tumours. J. Clin. Endocrinol. Metab. 57:1125.

Van Landegham A.A.J., Poortman J., Nabuurs M. and Thyssen J.H.H. 1985. Endogenous concentration and subcellular distribution of estrogens in normal and malignant breast tissue. Cancer Research 45:2900.

Vermeulen A. 1986. Human mammary cancer as a site of sex steroid metabolism. Cancer Surv. 5:585.

Vermeulen A., Deslypere J.P., Paridaens R., Leclerq C., Roy F. and Heuson J.C. 1986. Aromatase, 17β-hydroxysteroid dehydrogenase and intratissular sex hormone concentration in cancerous and normal glandular breast tissue in postmenopausal women. Eur. J. Cancer Clin. Oncol. 22:515.

INDEX

GPSR Compliance
The European Union's (EU) General Product Safety Regulation (GPSR) is a set of rules that requires consumer products to be safe and our obligations to ensure this.

If you have any concerns about our products, you can contact us on

ProductSafety@springernature.com

In case Publisher is established outside the EU, the EU authorized representative is:

Springer Nature Customer Service Center GmbH
Europaplatz 3
69115 Heidelberg, Germany

www.ingramcontent.com/pod-product-compliance
Ingram Content Group UK Ltd.
Pitfield, Milton Keynes, MK11 3LW, UK
UKHW051130260726
13967UKWH00010B/2961
* 9 7 8 1 4 6 8 4 5 7 5 2 0 *